Calvin Frank Allen

Railroad Curves and Earthwork

Calvin Frank Allen

Railroad Curves and Earthwork

ISBN/EAN: 9783744678858

Printed in Europe, USA, Canada, Australia, Japan

Cover: Foto ©berggeist007 / pixelio.de

More available books at **www.hansebooks.com**

RAILROAD CURVES

AND

EARTHWORK

BY

C. FRANK ALLEN, S.B.

MEMBER AMERICAN SOCIETY OF CIVIL ENGINEERS
PROFESSOR OF RAILROAD ENGINEERING IN THE MASSACHUSETTS
INSTITUTE OF TECHNOLOGY

NEW YORK
SPON & CHAMBERLAIN, 12 CORTLANDT STREET
LONDON
E. & F. N. SPON, Ltd., 125 STRAND

COPYRIGHT, 1889 AND 1894,
BY C. F. ALLEN.

TYPOGRAPHY BY J. S. CUSHING & CO., NORWOOD, MASS.

JOHN WILSON & SONS, CAMBRIDGE, MASS.

PREFACE.

This book was prepared for the use of the students in the author's classes. It has been used in lithographed sheets for a number of years in very nearly the present form, and has given satisfaction sufficient to suggest putting it in print. An effort has been made to have the demonstrations simple and direct, and special care has been given to the arrangement and the typography, in order to secure clearness and conciseness of mathematical statement. Much of the material in the earlier part of the book is necessarily similar to that found in one or more of several excellent field books, although the methods of demonstration are in many cases new. This will be found true especially in Compound Curves, for which simple treatment has been found quite possible. New material will be found in the chapters on Turnouts and on "Y" Tracks and Crossings. The Spiral Easement Curve is treated originally. The chapters on Earthwork are essentially new; they include Staking Out; Computation, directly and with Tables and Diagrams; also Haul, treated ordinarily and by Mass Diagram. Most of the material relating to Earthwork is not elsewhere readily available for students' use.

The book has been written especially to meet the needs of students in engineering colleges, but it is probable that it will be found useful by many engineers in practice. The size of page allows it to be used as a pocket book in the field. It is difficult to avoid typographical and clerical errors; the author will consider it a favor if he is notified of any errors found to exist.

C. FRANK ALLEN.

Boston,
September, 1899.

CONTENTS.

CHAPTER I.

Reconnoissance.

SECTION	PAGE
1. Operations in location	1
2. Reconnoissance	1
3. Nature of examination	1
4. Features of topography	2
5. Purposes of reconnoissance	3
6. Elevations, how taken	3
7. Pocket instruments used	4
8. Importance of reconnoissance	5

CHAPTER II.

Preliminary Survey.

9. Nature of preliminary	6
10. Grades	6
11. Importance of low grades	7
12. Pusher grades	7
13–14. Purposes of preliminary	8
15. Nature	9
16. Methods	9
17. Backing up	10
18. Notes	11
19. Organization of party	11
20. Locating engineer	11
21. Transitman; also form of notes	12
22. Head chainman	13
23. Stakeman	13
24. Rear chainman	14
25. Back flag	14
26. Axeman	14
27. Leveler; also form of notes	14

SECTION	PAGE
28. Rodman	15
29. Topographer	16
30. Preliminary by stadia	17

CHAPTER III.

Location Survey.

31. Nature of "location"	18
32. First method	18
33. Second method	18
34. Long tangents	19
35. Tangent from broken line of preliminary	19
36. Method of staking out location	19

CHAPTER IV.

Simple Curves.

37. Definitions	20
38. Measurements	20
39. Degree of curve	20
40–41. Formulas for degree and radius	21
42. Approximate method	21
43. Tangent distance T	22
44. Approximate method	22
45. External distance E	23
46. Middle ordinate M	23
47. Chord C	23
48. Formulas for R and D in terms of T, E, M, C, I	24
49. Sub-chord c	24
50. Sub-angle d	24
51. Length of curve L	26
52. Method of deflection angles	27
53. Example; application to parabola	27
54–55. Deflection angles for simple curves	27–28
56. Field-work	29
57. Example	30
58. Caution	30
59. Field-work when curve cannot be laid out from $P.C.$	31
60. Second method	31
61. Field-work of finding $P.C.$ and $P.T.$	32
62. Example	33

SECTION	PAGE
63. Form of transit book for curves	34
64. Curves with metric system	35
65. Method of deflection distances	36
66. Field-work	36
67. Approximate computation of offsets	37
68. Offset between two curves	37
69. Deflection distances for curve beginning with sub-chord	37
70. Example	38
71. Approximate solution of right triangles	38
72. Field-work for method of deflection distances	39
73. Caution	39
74. Deflection distances when first sub-chord is short	40
75. Method of offsets from the tangent	40
76. Field-work	40
77–78. Middle ordinates	41
79. Ordinate at any point	41
80. Middle ordinate, approximate formula	42
81. Example	42
82–83. Any ordinate, approximate formulas	42–43
84. Example	43
85. Find a series of points by middle ordinates	44
86–88. Substitute new curves to end in parallel tangents	44–46
89–90. Curve to join tangents and pass through given point	46–47
91–92. Find where curve and given line intersect	47–48
93. Approximate method	48
94. Find tangent from curve to given point	48
95. Approximate method	49
96–99. Obstacles in cases of curves	50–51

CHAPTER V.

Compound Curves.

100. Definitions	52
101. Field-work	52
102. Data	53
103. Given R_l, R_s, I_l, I_s; required I, T_l, T_s	53
104. Given T_s, R_s, R_l, I; required I_s, I_l, T_l	54
105. Given T_s, R_s, I_s, I; required I_l, T_l, R_l	54
106. Given T_l, T_s, R_s, I; required I_l, I_s, R_l	54
107. Given T_l, R_l, R_s, I; required I_s, I_l, T_s	55
108. Given T_l, R_l, I_l, I; required I_s, T_s, R_s	55
109. Given T_l, T_s, R_l, I; required I_s, I_l, R_s	55

SECTION	PAGE
110. Given, long chord, angles, and R_s; required I_l, I_s, I, R_l.	56
111. Given, long chord, angles, and R_l; required I_l, I_s, I, R_s.	56
112. Substitute for simple, a compound curve to end in parallel tangent	56
113. Example	57
114. Change $P.C.C.$ so as to join parallel tangent	58
115. Substitute for simple, a symmetrical curve with flattened ends	59
116. Substitute curve with flattened ends to pass through middle point	60
117. Substitute simple curve for curves with connecting tangent	61

CHAPTER VI.

Reversed Curves.

118. Use of reversed curves	62
119–122. Between parallel tangents, common radius	62–63
123–124. Between parallel tangents, unequal radii	63–64
125–126. Find common radius to connect tangents not parallel	64–65
127–128. Find I_1, I_2, T_2 when I, T_1, R_1, R_2 are given	65–66

CHAPTER VII.

Parabolic Curves.

129. Use of parabolic curves	67
130. Properties of the parabola	67
131. Lay out parabola by offsets from tangent	68
132. Field-work	69
133. Parabola by middle ordinates	70
134. Vertical curves, where used	70
135. Method for vertical curve 200 feet long	71
136. General method	72
137. Example	74
138. To find proper length of vertical curve	74

CHAPTER VIII.

Turnouts.

139. Definitions	75
140. Find frog angle from number of frog	76
141–142. Descriptions of features of turnouts	77

Contents.

SECTION		PAGE
143.	Given, gauge, frog angle, and throw; required, radius, length, and lead	78
144.	Given, gauge, number, and throw; required, radius, length, and lead	79
145.	Given, gauge; required, middle ordinate	79
146.	Find angle of crotch frog	80
147.	Find number of crotch frog	80
148.	Find proper radius for turnout inside; also lead	81
149.	Approximate formula to find degree of turnout inside	82
150–151.	Find radius of turnout outside; also lead	83–84
152.	Example of case, § 148	84
153.	Description of split switch	85
154.	Find radius of turnout for split switch	86
155.	Complication on account of straight frog	87
156.	Find radius of turnout for split switch and straight frog	87
157.	Methods of connecting parallel tracks by turnouts	88
158–162.	Formulas for these	88–89
163.	Formulas for a series of parallel tracks	90
164.	Find radius of turnout curve from frog to parallel curved track outside	90
165.	Approximate method	91
166.	Example. Precise method	92
167.	Approximate method	92
168.	Find radius of turnout curve from frog to parallel curved track inside	93
169.	Special case for turnout outside	93
170.	Calculations for turnout between parallel curved tracks	94
171.	Approximate method	94

CHAPTER IX.

"Y" Tracks and Crossings.

172.	Definition	95
173.	Main track tangent, "Y" track curved, and turnout curved	95
174.	Main track tangent, "Y" curved, turnout curved with tangent	96
175–176.	Main track tangent and curve "Y" curved, turnout curved	96–97
177.	Crossing of tangent and curve	97
178.	Crossing of two curves	98

CHAPTER X.

Spiral Easement Curve.

SECTION	PAGE
179. Necessity for; also elevation of outer rail.	99
180. Simple equation for curve.	100
181. Equation of cubic parabola derived	100
182. Discussion of character of easement curves.	101
183. Field-work for cubic parabola	102
184. Statement of requisites of curve described	103
185-188. Demonstrations	103-105
189. Field-work by deflection angles.	106
190-191. Requisites of curve described, and demonstrations.	106-107
192. Character of approximations used.	107
193. Cubic parabola to connect the parts of compound curve.	108
194. Cubic parabola for compound curves by deflection angles	108
195. To connect parts of compound curve by Searles' Spiral.	109
196. Example	110

CHAPTER XI.

Setting Stakes for Earthwork.

197. Data	111
198. What stakes and how marked	111
199. Method of finding rod reading for grade	112
200. Example	113
201. Cut or fill at center.	113
202. Side stake for level section	114
203-206. Side stakes when surface is not level.	114-116
207. Slope-board or level-board	116
208-210. Keeping the notes	117
211-212. Form of note-book.	118-119
213. Cross-sections; where taken	120
214-215. Passing from cut to fill	120-121
216. Opening in embankment.	121
217. General level notes	121
218-221. Level, three-level, five-level, irregular, sections	122

CHAPTER XII.

Methods of computing Earthwork.

222. Principal methods used	123
223. Averaging end areas.	123

Contents. xi

SECTION	PAGE
224. Kinds of cross-sections specified	124
225. Level cross-section	124
226. Three-level section	125
227. Three-level section; second method	126
228. Five-level section	127
229. Irregular section	127
230. Planimeter	128
231. Comment on end area formula	128
232. Prismoidal formula	128
233. Prismoidal formula for prisms, wedges, pyramids	129
234. Nature of regular section of earthwork	130
235–237. Prismoidal formula applied where upper surface is warped	130–132
238. Wide application of prismoidal formula	132
239–240. Prismoidal correction	133–134
241. Where applicable; also special case	135
242. Correction for pyramid	136
243. Correction for five-level sections	136
244–245. Correction for irregular sections	136–137
246. Value of prismoidal correction	137
247. Method of middle areas	138
248. Method of equivalent level sections	138
249. Method of mean proportionals	138
250. Henck's method	138
251. Formula	139
252–254. Example	140–141
255–256. Comment on Henck's and end area methods	141–142
257–263. Examples comparing the various methods	142–144

CHAPTER XIII.

SPECIAL PROBLEMS IN EARTHWORK.

264. Correction for curvature	145
265. Correction where chords are less than 100 ft.	147
266. Correction of irregular sections	147
267. Opening in embankment	148
268. Comment	150
269. Borrow-pits	150
270. Truncated triangular prism	150
271. Truncated rectangular prism	151
272. Assembled prisms	153

CHAPTER XIV.

EARTHWORK TABLES.

SECTION	PAGE
273. Formula for use in tables.	154
274. Arrangement of table.	155
275. Explanation of table.	155
276-277. Example of use, including prismoidal correction table	156
278. Prismoidal correction applied for section less than 100 feet	157
279. Tables, where published	157
280. Tables of triangular prisms.	157
281. Where published	157
282. Arrangement of tables of triangular prisms	158
283. Example of use.	159
284-285. Application to irregular sections.	160

CHAPTER XV.

EARTHWORK DIAGRAMS.

286-287. Method of diagrams	161-162
288. Forms of equations available for straight lines	162
289. Method of use of diagrams.	162
290-291. Computations and table for diagram of prismoidal correction	163-164
292. Diagram for prismoidal correction and explanation of construction	164-165
293. Explanation and example of use.	166
294. Table of triangular prisms.	166
295-298. Computations and table for diagram of three-level sections	167-169
299-300. Checks upon computations.	171
301. Explanation of diagram; also curve of level section	171
302. Use of diagram for three-level sections	172
303. Comment on rapidity by use of diagrams.	173
304. Special use to find prismoidal correction for irregular sections	173

CHAPTER XVI.

HAUL.

305. Definition and measure of haul.	174
306-307. Length of haul, how found	174-175
308. Formula for center of gravity of a section.	175

Contents. xiii

SECTION	PAGE
309–310. Formula deduced	176–177
311. Formula modified for use with tables or diagrams	178
312. For section less than 100 feet	178
313. For series of sections	179

CHAPTER XVII.

Mass Diagram.

314. Definition	180
315. Table and method of computation	180–181
316. Mass diagram and its properties	182–183
317. Graphical measure of haul explained	183
318. Application to mass diagram	184–185
319. Further properties	185
320. Mass diagram; showing also borrow and waste	186–187
321. Profitable length of haul	187
322–323. Example of use of diagram	188–189

Tables and Diagrams 190–202

RAILROAD CURVES AND EARTHWORK.

CHAPTER I.

1. The operations of "locating" a railroad, as commonly practiced in this country, are three in number: —

    I. RECONNOISSANCE.
    II. PRELIMINARY SURVEY.
    III. LOCATION SURVEY.

I. RECONNOISSANCE.

2. **The Reconnoissance** is a rapid survey, or rather a critical examination of country, without the use of the ordinary instruments of surveying. Certain instruments, however, are used, the Aneroid Barometer, for instance. It is very commonly the case that the termini of the railroad are fixed, and often intermediate points also. It is desirable that no unnecessary restrictions as to intermediate points should be imposed on the engineer to prevent his selecting what he finds to be the best line, and for this reason it is advisable that the reconnoissance should, where possible, precede the drawing of the charter.

3. The first step in reconnoissance should be to procure the best available maps of the country; a study of these will generally furnish to the engineer a guide as to the routes or section of country that should be examined. If maps of the United States Geological Survey are at hand, with contour lines and other topography carefully shown, the reconnoissance can be largely determined upon these maps. Lines clearly impracticable will be thrown out, the maximum grade closely determined, and the field examinations reduced to a minimum. No

route should be accepted finally from any such map, but a careful field examination should be made over the routes indicated on the contour maps. The examination, in general, should cover the general section of country, rather than be confined to a single line between the termini. A straight line and a straight grade from one terminus to the other is desirable, but this is seldom possible, and is in general far from possible. If a single line only is examined, and this is found to be nearly straight throughout, and with satisfactory grades, it may be thought unnecessary to carry the examination further. It will frequently, however, be found advantageous to deviate considerably from a straight line in order to secure satisfactory grades. In many cases it will be necessary to wind about more or less through the country in order to secure the best line. Where a high hill or a mountain lies directly between the points, it may be expected that a line around the hill, and somewhat remote from a direct line, will prove more favorable than any other. Unless a reasonably direct line is found, the examination, to be satisfactory, should embrace all the section of intervening country, and all feasible lines should be examined.

4. There are two features of topography that are likely to prove of especial interest in reconnoissance, *ridge lines* and *valley lines*.

A *ridge line* along the whole of its course is higher than the ground immediately adjacent to it on each side. That is, the ground slopes downward from it to both sides. It is also called a *watershed line*.

A *valley line*, to the contrary, is lower than the ground immediately adjacent to it on each side. The ground slopes upward from it to both sides. Valley lines may be called *watercourse lines*.

A *pass* is a place on a ridge line lower than any neighboring points on the same ridge. Very important points to be determined in reconnoissance are the passes where the ridge lines are to be crossed; also the points where the valleys are to be crossed; and careful attention should be given to these points. In crossing a valley through which a large stream flows, it may be of great importance to find a good bridge crossing. In some cases where there are serious difficulties in crossing a ridge, a tunnel may be necessary. Where such structures, either bridges or tunnels, are to be built, favorable points for their construction should be selected and the rest of the line be compelled to conform. In many parts of the United States at the present time, the necessity for avoiding grade crossings causes the crossings of roads and streets to become governing points of as great importance as ridges and valleys.

5. There are several purposes of reconnoissance: first, to find whether there is any satisfactory line between the proposed termini; if so, second, to establish which is the most feasible; third, to determine approximately the maximum grade necessary to be used; fourth, to report upon the character or geological formation of the country, and the probable cost of construction depending somewhat upon that; fifth, to make note of the existing resources of the country, its manufactures, mines, agricultural or natural products, and the capabilities for improvement and development of the country resulting from the introduction of the railroad. The report upon reconnoissance should include information upon all these points. It is for the determination of the third point mentioned, the rate of maximum grade, that the barometer is used. Observing the elevations of governing points, and knowing the distances between those points, it is possible to form a good judgment as to what rate of maximum grade to assume.

6. The Elevations are usually taken by the *Aneroid Barometer*. Tables for converting barometer readings into elevations above sea-level are readily available and in convenient form for field use. (See Searles' or Henck's Field Books.)

Distances may be determined with sufficient accuracy in many cases from the map, where a good one exists. Where this method is impossible or seems undesirable, the distance may be determined in one of several different ways. When the trip is made by wagon, it is customary to use an *Odometer*, an instrument which measures and records the number of revolutions of the wheel to which it is attached, and thus the distance traveled by the wagon. There are different forms of odometer. In its most common form, it depends upon a hanging weight or pendulum, which is supposed to hold its position, hanging vertical, while the wheel turns. The instrument is attached to the wheel between the spokes and as near to the hub as practicable. At low speeds it registers accurately; as the speed is increased, a point is reached where the centrifugal force neutralizes or overcomes the force of gravity upon the pendulum, and the instrument fails to register accurately, or perhaps at high speeds to register at all. If this form of odometer is used, a clear understanding should be had of the conditions under which it fails to correctly register. A theoretical discussion might closely establish the point at which the centrifugal force will balance the force of gravity. The wheel striking against stones in a rough road will create disturbances in the action of the pendulum, so that the odometer will fail to register accurately at speeds less than that determined upon the above assumption.

Another form of odometer is manufactured which is connected both with the wheel and the axle, and so measures positively the relative motion between the wheel and axle, and this ought to be reliable for registering accurately. Many engineers prefer to count the revolutions of the wheel themselves, tying a rag to the wheel to make a conspicuous mark for counting.

When the trip is made on foot, pacing will give satisfactory results. An instrument called the *Pedometer* registers the results of pacing. As ordinarily constructed, the graduations read to quarter miles, and it is possible to estimate to one-tenth that distance. Pedometers are also made which register paces. In principle, the pedometer depends upon the fact that, with each step, a certain shock or jar is produced as the heel strikes the ground, and each shock causes the instrument to register. Those registering miles are adjustable to the length of pace of the wearer.

If the trip is made on horseback, it is found possible to get good results with a steady-gaited horse, by first determining his rate of travel and figuring distance by the time consumed in traveling. Excellent results are said to have been secured in this way.

7. It is customary for engineers not to use a compass in reconnoissance, although this is sometimes done in order to trace the line traversed upon the map, and with greater accuracy. A pocket level will be found useful. The skillful use of pocket instruments will almost certainly be found of great value to the engineer of reconnoissance.

It may, in cases, occur that no maps of any value are in existence or procurable. It may be necessary, in such a case, to make a rapid instrumental survey, the measurements being taken either by pacing, chain, or stadia measurements. This is, however, unusual.

8. The preliminary survey is based upon the results of the reconnoissance, and the location upon the results of the preliminary survey. The reconnoissance thus forms the foundation upon which the location is made. Any failure to find a suitable line and the best line constitutes a defect which no amount of faithfulness in the later work will rectify. The most serious errors of location are liable to be due to imperfect reconnoissance; an inefficient engineer of reconnoissance should be avoided at all hazards. In the case of a new railroad, it would, in general, be proper that the Chief Engineer should in person conduct this survey. In the case of the extension of existing lines, this might be impracticable or inadvisable, but an assistant of known responsibility, ability, and experience should in this case be selected to attend to the work.

CHAPTER II.

II. PRELIMINARY SURVEY.

9. **The Preliminary Survey** is based upon the results of the reconnoissance. It is a survey made with the ordinary instruments of surveying. Its purpose is to fix and mark upon the ground a first trial line approximating as closely to the proper final line as the difficulty of the country and the experience of the engineer will allow; further than this, to collect data such that this survey shall serve as a basis upon which the final Location may intelligently be made. In order to approximate closely in the trial line, it is essential that the maximum grade should be determined or estimated as correctly as possible, and the line fixed with due regard thereto.

It will be of value to devote some attention here to an explanation about Grades and "Maximum Grades."

10. Grades. — The ideal line in railroad location is a straight and level line. This is seldom, if ever, realized. When the two termini are at different elevations, a line straight and of uniform grade becomes the ideal. It is commonly impossible to secure a line of uniform grade between termini. In operating a railroad, an engine division will be about 100 miles, sometimes less, often more. In locating any 100 miles of railroad, it is almost certain that a uniform grade cannot be maintained. More commonly there will be a succession of hills, part of the line up grade, part down grade. Sometimes there will be a continuous up grade, but not at a uniform rate. With a uniform grade, a locomotive engine will be constantly exerting its maximum pull or doing its maximum work in hauling the longest train it is capable of hauling; there will be no power wasted in hauling a light train over low or level grades upon which a heavier train could be hauled. Where the grades are not uniform, but are rising or falling, or rising irregularly, it will be found that the topography on some particular 5 or 10 miles is of such a character that the grade here must be steeper than is really necessary anywhere else on the line; or there may be two or three stretches of grade where about the same rate of grade is necessary, steeper than elsewhere required. The steep grade thus found necessary at some special point or points on the line of railroad is called the "Maximum Grade" or "Ruling Grade" or "Limiting Grade," it being the grade that limits the weight of train that an engine can haul over the whole division. It should then be the effort to make the rate of maximum grade as low as possible, because the lower the rate of the maximum grade, the heavier the train a given locomotive can haul, and because it costs not very much more to haul a heavy train than a light one. The maximum grade determined by the reconnoissance should be used as the basis for the preliminary survey. How will this affect the line? Whenever a hill is encountered, if the maximum grade be steep, it may be possible to carry the line straight, and over the hill; if the maximum grade be low, it may be necessary to deflect the line and carry it around the hill. When the maximum grade has been once properly determined, if any saving can be accomplished by using it rather than a grade less steep, the maximum grade should be used. It is possible that the train loads will not be uniform throughout the division. It will be advantageous to spend a small sum of money to keep any grade lower than the maximum, in view of the *possibility* that at this particular point the train load will be heavier than elsewhere on the division. Any saving made will in general be of one or more of three kinds:—

a. Amount or "quantity" of excavation or embankment;
b. Distance;
c. Curvature.

11. In some cases, a satisfactory grade, a low grade for a maximum, can be maintained throughout a division of 100 miles in length, with the exception of 2 or 3 miles at one point only. So great is the value of a low maximum grade that all kinds of expedients will be sought for, to pass the difficulty without increasing the rate of maximum grade, which we know will apply to the whole division.

12. Sometimes by increasing the length of line, we are able to reach a given elevation with a lower rate of grade. Some-

times heavy and expensive cuts and fills may serve the purpose. Sometimes all such devices fail, and there still remains necessary an increase of grade at this one point, but at this point only. In such case it is now customary to adopt the higher rate of grade for these 2 or 3 miles and operate them by using an extra or additional engine. In this case, the "ruling grade" for the division of 100 miles is properly the "maximum grade" prevailing over the division generally, the higher grade for a few miles only being known as an "Auxiliary Grade" or more commonly a "Pusher Grade." The train which is hauled over the engine division is helped over the auxiliary or pusher grade by the use of an additional engine called a "Pusher." Where the use of a short "Pusher Grade" will allow the use of a low "maximum grade," there is evident economy in its use. The critical discussion of the importance or value of saving distance, curvature, rise and fall, and maximum grade, is not within the scope of this book, and the reader is referred to Wellington's "Economic Theory of Railway Location."

13. The Preliminary Survey follows the general line marked out by the reconnoissance, but this rapid examination of country may not have fully determined which of two or more lines is the best, the advantages may be so nearly balanced. In this case two or more preliminary surveys must be made for comparison. When the reconnoissance has fully determined the general route, certain details are still left for the preliminary survey to determine. It may be necessary to run two lines, one on each side of a small stream, and possibly a line crossing it several times. The reconnoissance would often fail to settle minor points like this. It is desirable that the preliminary survey should closely approximate to the final line, but it is not important that it should fully coincide anywhere.

An important purpose of the "preliminary" is to provide a map which shall show enough of the topography of the country, so that the Location proper may be projected upon this map. Working from the line of survey as a base line, measurements should be taken sufficient to show streams and various natural objects as well as the contours of the surface.

14. The Preliminary Survey serves several purposes: —

First. To fix accurately the maximum grade for use in Location.

Second. To determine which of several lines is best.

Third. To provide a map as a basis upon which the Location can properly be made.

Fourth. To make a close estimate of the cost of the work.

Fifth. To secure, in certain cases, legal rights by filing plans.

15. It should be understood that the preliminary survey is, in general, simply a means to an end, and rapidity and economy are desirable. It is an instrumental survey. Measurements of distance are taken usually with the chain, although a tape is sometimes used. Angles are taken generally with a transit; some advocate the use of a compass. The line is ordinarily run as a broken line with angles, but is occasionally run with curves connecting the straight stretches, generally for the reason that a map of such a line is available for filing, and certain legal rights result from such a filing. With a compass, no backsight need be taken, and, in passing small obstacles, a compass will save time on this account. A transit line can be carried past an obstacle readily by a zigzag line. Common practice among engineers favors the use of the transit rather than the compass. Stakes are set, at every "Station," 100 feet apart, and the stakes are marked on the face, the first 0, the next 1, then 2, and so to the end of the line. A stake set 1025 feet from the beginning would be marked 10 + 25.

Levels are taken on the ground at the side of the stakes, and as much oftener as there is any change in the inclination of the ground. All the surface heights are platted on a profile, and the grade line adjusted.

16. The line should be run from a governing point towards country allowing a choice of location, that is from a pass or from an important bridge crossing, towards country offering no great difficulties. There is an advantage in running from a summit downhill, subject, however, to the above considerations. In running from a summit down at a prescribed rate of grade, an experienced engineer will carry the line so that, at the end of a day's work, the levels will show the line to be about where it ought to be. For this purpose, the levels must be worked up and the profile platted to date at the close of each day. Any slight change of line found necessary can then be made early the next morning. A method sometimes adopted in working down from a summit is for the locating engineer to plat his grade line on the profile, daily in advance, and then during the day, plat a point on his profile whenever he can conveniently get one from his leveler, and thus find whether his line is too high or too low.

17. Occasionally the result of two or three days' work will yield a line extremely unsatisfactory, enough so that the work of these two or three days will be abandoned. The party "backs up" and takes a fresh start from some convenient point. In such case the custom is not to tear out several pages of note-book, but instead to simply draw a line across the page and mark the page "Abandoned." At some future time the abandoned notes may convey useful information to the effect that this line was attempted and found unavailable. In general, all notes worth taking are worth saving.

Sometimes after a line has been run through a section of country, there is later found a shorter or better line.

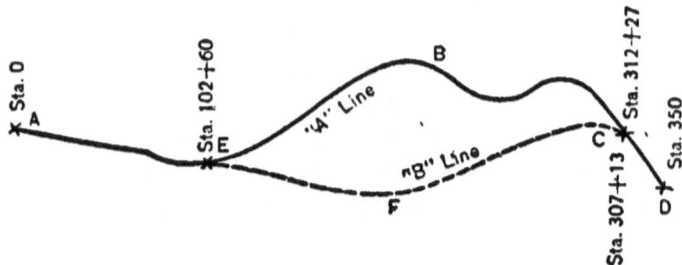

In the figure used for illustration, the first line, "A" Line, is represented by AEBCD, upon which the stations are marked continuously from A to D, 350 stations. The new line, "B" Line, starts from E, Sta. 102 + 60, and the stationing is held continuous from 0 to where it connects with the "A" Line at C. The point C is Sta. 312 + 27 of the "A" Line, and is also Sta. 307 + 13 of the "B" Line. It is not customary to restake the line from C to D in accordance with "B" Line stationing. Instead of this, a note is made in the note-books as follows: —

   Sta. 312 + 27 "A" Line = 307 + 13 "B" Line.

Some engineers make the note in the following form: —

   Sta. 307 to 313 = 86 ft.

The first form is preferable, being more direct and less liable to cause confusion.

18. All notes should be kept clearly and nicely in a note-book — never on small pieces of paper. The date and the names of members of the party should be entered each day in the upper left-hand corner of the page. An office copy should be made as soon as opportunity offers, both for safety and convenience. *The original notes should always be preserved;* they would be admissible as evidence in a court of law where a copy would be rejected. When two or more separate or alternate lines are run, they may be designated

Line " A," Line " B," Line " C,"

or " A " Line, " B " Line, " C " Line.

19. The Organization of Party may be as follows : —

1. Locating Engineer.
2. Transitman. ⎫
3. Head Chainman. ⎪
4. Stakeman. ⎬ Transit Party.
5. Rear Chainman. ⎪
6. Back Flag. ⎪
7. Axemen (one or more). ⎭
8. Leveler. ⎫
9. Rodman (sometimes two). ⎬ Level Party.
10. Topographer. ⎫
11. Assistant. ⎬ Topographical Party
12. Cook.
13. Teamster.

20. The Locating Engineer is the chief of party, and is responsible for the business management of the camp and party, as well as for the conduct of the survey. He determines where the line shall run, keeping ahead of the transit, and establishing points as foresights or turning-points for the transitman. In open country, the extra axeman can assist by holding the flag at turning-points, and thus allowing the locating engineer to push on and pick out other points in advance. The locating engineer keeps a special note-book or memorandum book ; in it he notes on the ground the quality of material, rock, earth, or whatever it may be ; takes notes to determine the lengths and positions of bridges, culverts, and other structures ; shows the localities of timber, building stones, borrow pits, and

12 *Railroad Curves and Earthwork.*

other materials valuable for the execution of the work; in fact, makes notes of all matters not properly attended to by the transit, leveling, or topography party. The rapid and faithful prosecution of the work depend upon the locating engineer, and the party ought to derive inspiration from the energy and vigor of their chief, who should be the leader in the work. In open and easy country, the locating engineer may instill life into the party by himself taking the place of the head chainman occasionally. In country of some difficulty, his time will be far better employed in prospecting for the best line.

21. The Transitman does the transit work, ranges in the line from the instrument, measures the angles, and keeps the notes of the transit survey. The following is a good form for the left-hand page of the note-book:—

Station	Point	Deflection	Observed Bearing	Calculated Bearing
7			N 3° 30′ E	N 3° 38′ E
6	⊙ + 24	33° 02′ R		
5			N 29° 30′ W	N 29° 24′ W
4	⊙	12° 09′ L		
3				
2				
1			N 17° 15′ W	N 17° 15′ W
0	⊙			

Notes of topography and remarks are entered on the right-hand page, which, for convenience, is divided into small squares by blue lines, with a red line running up and down through the middle.

The stations run from bottom to top of page. The bearing is taken at each setting and recorded *just above* the corresponding point in the note-book, or opposite a part of the line, rather than opposite the point. Ordinarily, the transitman takes the bearings of all fences and roads crossed by the line, finds the stations from the rear chainman, and records them in their proper place and direction on the right-hand page of the note-book. Section lines of the United States Land Surveys should be observed in the same way. The transitman is next in authority to the locating engineer, and directs the work when the latter is not immediately present. The transitman, while moving from point to point, setting up, and ranging line, limits the speed of the entire party, and should waste no time.

22. The Head Chainman carries a "flag" and the forward end of the chain, which should be held level and firm with one hand, while the flag is moved into line with the other. He should always put himself nearly in line before receiving a signal from the transitman; plumbing may be done with the flag. When the point is found, the stakeman will set the stake. When a suitable place for a turning-point is reached, a signal should be given the transitman to that effect. A nail should be set in top of the stake at all turning-points. A proper understanding should be had with the transitman as to signals.

Signals from the Transitman.

A horizontal movement of the hand indicates that the rod should be moved as directed.

A swinging movement of the hand, "Plumb the rod as indicated."

A movement of both hands, or waving the handkerchief freely above the head, means "All right."

At long distances, a handkerchief can be seen to advantage; when snow is on the ground, something black is better.

Signals from the Head Chainman.

Setting the bottom of flag on the ground and waving the top, means "Give the line."

Raising the flag above the head and holding it horizontal with both hands: "Give line for a turning-point."

The "all right" signal is the same as from the transitman.

In all measurements less than 100 feet (or a full chain), the head chainman holds the end of the chain, leaving the reading of the measurement to the rear chainman.

The head chainman regulates the speed of the party during the time that the instrument is in place, and should keep alive all the time. The rear chainman will keep up as a matter of necessity.

23. The Stakeman carries, marks, and drives the stakes at the points indicated by the head chainman. The stakes should be driven with the flat side towards the instrument, and marked on the front with the number of the station. Intermediate stakes should be marked with the number of the last station + the additional distance in feet and tenths, as 10 + 67.4. The stationing is not interrupted and taken up anew at each turning-point, but is continuous from beginning to end of the survey. At each turning-point a plug should be driven nearly flush with the ground, and a witness stake driven, in an inclined position, at a distance of about 15 inches from the plug, and at the side towards which the advance line deflects, and marked W and under it the station of the plug.

24. The Rear Chainman holds the rear end of the chain over the stake last set, but does not hold against the stake to loosen it. He calls "Chain" each time when the new stake is reached, being careful not to overstep the distance. He should stand beside the line (not on it) when measuring, and take pains not to obstruct the view of the transitman. He checks, and is responsible for the correct numbering of stakes, and for all distances less than 100 feet, as the head chainman always holds the end of the chain. The stations where the line crosses fences, roads, and streams should be set down in a small note-book, and reported to the transitman at the earliest convenient opportunity. The rear chainman is responsible for the chain.

25. The Back Flag holds the flag as a backsight at the point last occupied by the transit. The only signals necessary for him to understand from the transitman are "plumb the flag" and "all right." The flag should always be in position, and the transitman should not be delayed an instant. The back flag should be ready to come up the instant he receives the "all right" signal from the transitman. The duties are simple, but frequently are not well performed.

26. The Axeman cuts and clears through forest or brush. A good axeman should be able to keep the line well, so as to cut nothing unnecessary. In open country, he prepares the stakes ready for the stakeman or assists the locating engineer as *fore flag*.

27. The Leveler handles the level and generally keeps the notes, which may have the following form for the left-hand page. The right-hand page is for remarks and descriptions of turning-points and bench-marks. It is desirable that turning-

Preliminary Survey. 15

Station	+ S	H I	– S	Elevation
B.M.	4.67	104.67		100.00
0			5.7	99.0
1			6.9	97.8
2			3.4	101.3
T.P.	9.26	112.81	1.12	103.55
3			8.5	104.3

points should, where possible, be described, and that all benchmarks should be used as turning-points. Readings on turning-points should be recorded to hundredths or to thousandths of a foot, dependent upon the judgment of the Chief Engineer. Surface readings should be made to the nearest tenth, and elevations set down to nearest tenth only. A self-reading rod has advantages over a target rod for short sights. A target rod is possibly better for long sights and for turning-points. The "Philadelphia Rod" is both a target rod and a self-reading rod, and is thus well adapted for railroad use. Bench-marks should be taken at distances of from 1000 to 1500 feet, depending upon the country. All bench-marks, as soon as calculated, should be entered together on a special page near the end of the book. The leveler should test his level frequently to see that it is in adjustment. The leveler and rodman should together bring the notes to date every evening and plat the profile to correspond.

The profile of the preliminary line should show : —

a. Surface line (in black).
b. Grade line (in red).
c. Grade elevations at each change in grade (in red).
d. Rate of grade, per 100 (in red); rise +, fall –.
e. Station and deflection at each angle in the line (in black).
f. Notes of roads, ditches, streams, bridges, etc. (in black).

28. The Rodman carries the rod and holds it vertical upon the ground at each station and at such intermediate points as mark any important change of slope of the ground. The surface of streams and ponds should be taken when met, and at frequent intervals where possible, if they continue near the line.

Levels should also be taken of high-water marks wherever traces of these are visible. The rodman carries a small notebook in which he enters the rod readings at all turning-points. In country which is open, but not level, the transit party is liable to outrun the level party. In such cases greater speed will be secured by the use of two rodmen.

29. The Topographer is, or should be, one of the most valuable members of the party. In times past it has not always been found necessary to have a topographer, or if employed, his duty has been to sketch in the general features necessary to make an attractive map, and represent hills and buildings sufficiently well with reference to the line to show, in a general way, the reason for the location adopted. Sometimes the chief of the party has for this purpose taken the topography. At present the best practice favors the taking of accurate data by the topography party.

The topographer (with one or two assistants) should take the station and bearing (or angle) of every fence or street line crossed by the survey (unless taken by the transit party); also take measurements and bearings for platting all fences and buildings near enough to influence the position of the Location; also sketch, as well as may be, fences, buildings, and other topographical features of interest which are too remote to require exact measurement; and finally, establish the position of contour lines, streams, and ponds, within limits such that the Location may be well and fully determined in the contoured map.

The work of taking contours is accomplished by the use of hand level and tape (pacing may, in many cases, be sufficient). The elevation at the center line at each station is found from the leveler. Contours are generally taken at vertical intervals of 5 feet. The contour line, say to the right of the station, is found by reading upon a light hand-marked rod the difference in elevation between the center line at the stake, and the required contour. A cloth tape, held by the assistant topographer, will serve the purpose of a rod. The next contour to the right can be readily found, owing to the fact that the topographer's eye is nearly 5 feet above the ground, making leveling easy. An allowance should be made for the difference between 5 feet and the height of the eye. Sections both right and left should be taken as often as necessary, the distances to each contour measured, and the lines between the points thus determined sketched in on the ground. Books of convenient size are made and divided into small cross-sections to facilitate sketching. Cross-section blocks or pads will be considered equally good by some engineers. The distance to which contour lines should be taken depends on the character of the country. The object should be to take contours as far from the line as is necessary in order to furnish contours requisite for determining the position of the located line.

Instead of a hand level, some engineers use a clinometer and take and record side slopes.

Topography can be taken rapidly and well by stadia survey or by plane table. This is seldom done, as most engineers are not sufficiently familiar with their use.

30. Some engineers advocate making a general topographical survey of the route by stadia, instead of the survey above described. In this case no staking out by "stations" would be done. All points occupied by the transit should be marked by plugs, which can be used to aid in marking the Location on the ground after it is determined on the contour map. This method has been used a number of times, and is claimed to give economical and satisfactory results; it is probable that it will have constantly increasing use in the future, and will prove the best method in a large share of cases.

CHAPTER III.

III. LOCATION SURVEY.

31. The Location Survey is the final fitting of the line to the ground. In Location, curves are used to connect the straight lines or "tangents," and the alignment is laid out complete, ready for construction.

The party is much the same as in the preliminary, and the duties substantially the same. More work devolves upon the transitman on account of the curves, and more skill is useful in the head chainman in putting himself in position on curves. He can readily range himself on tangent. The form of notes will be shown later. The profile is the same, except that it shows, for alignment notes, the P.C. and P.T. of curves, and also the degree and central angle, and whether to the right or left.

It is well to connect frequently location stakes with preliminary stakes, when convenient, as a check on the work.

In making the location survey, two distinct methods are in use among engineers : —

32. First Method of Location. — Use preliminary survey and preliminary profile as guides in reading the country, and locate the line upon the ground. Experience in such work will enable an engineer to get very satisfactory results in this way, in nearly all cases. The best engineers, in locating in this way, as a rule lay the tangents first, and connect the curves afterwards. It will appear later how this is done.

33. Second Method. — Use preliminary line, preliminary profile, and especially the contour lines on the preliminary map; make a paper location, and run this in on the ground. Some go so far as to give their locating engineer a complete set of notes to run by. This is going too far. Whether it is best to go farther than to fix, on the map, the location of tangents, and specify the degree of curve, is a question. A conservative method is to do no more than this, and in some cases, leave the degree of curve even an open question. The second method is gaining in favor, but the first method is, even now, much used. It is well accepted, among engineers, that no reversed curve should be used; 200 feet of tangent, at least, should intervene. Neither should any curve be very short, say less than 300 feet in length.

34. A most difficult matter is the laying of a long tangent, so that it shall be straight. Lack of perfect adjustment and construction of instrument will cause a "swing" in the tangent. The best way is to run for a distant foresight. Another way is to have the transit as well adjusted as possible, and even then change ends every time in reversing, so that errors shall not accumulate. It will be noticed that the preliminary is run in without curves because more economical in time; sometimes curves are run however, either because the line can be run closer to its proper position, or sometimes in order to allow of filing plans with the United States or separate States.

35. In Location, a single tangent often takes the place of a broken line in the preliminary, and it becomes important to determine the direction of the tangent with reference to some part of the broken line. This is readily done by finding the coördinates of any given point with reference to that part of the broken line assumed temporarily as a meridian. The course of each line is calculated, and the coördinates of any point thus found. It simplifies the calculation to use some part of the preliminary as an assumed meridian, rather than to use the actual bearings of the lines. The coördinates of two points on the proposed tangent allow the direction of the tangent to be determined with reference to any part of the preliminary. When the angles are small, an approximation sufficiently close will be secured, by assuming in all cases that the cosine of the angle is 1.000000 and that the sines are directly proportional to the angles themselves. In addition to this, take the distances at the nearest even foot, and the calculation becomes much simplified.

36. The located line, or "Location," as it is often called, is staked out ordinarily by center stakes which mark a succession of straight lines, connected by curves to which the straight lines are tangent. The straight lines are by general usage called "**Tangents.**"

CHAPTER IV.

SIMPLE CURVES.

37. The curves most generally in use are circular curves, although parabolic and other curves are sometimes used. Circular curves may be classed as **Simple, Compound, Reversed,** or **Spiral.**

A **Simple Curve** is a circular arc, extending from one tangent to the next. The point where the curve leaves the first tangent is called the "$P.C.$," meaning the point of curvature, and the point where the curve joins the second tangent is called the "$P.T.$," meaning the point of tangency. The $P.C.$ and $P.T.$ are often called the Tangent Points. If the tangents be produced, they will meet in a point of intersection called the "**Vertex**," V. The distance from the vertex to the $P.C.$ or $P.T.$ is called the "**Tangent Distance**," T. The distance from the vertex to the curve (measured towards the center) is called the **External Distance**, E. The line joining the middle of the **Chord**, C, with the middle of the curve subtended by this chord, is called the **Middle Ordinate**, M. The radius of the curve is called the **Radius**, R. The angle of deflection between the tangents is called the **Intersection Angle**, I. The angle at the center subtended by a chord of 100 feet is called the **Degree of Curve**, D. A chord of less than 100 feet is called a **sub-chord**, c; its central angle a **sub-angle**, d.

38. The measurements on a curve are made:

(*a*) from $P.C.$ by a sub-chord (sometimes a full chord of 100 ft.) to the next even station, then

(*b*) by chords of 100 feet each between even stations, and finally,

(*c*) from the last station on the curve, by a sub-chord (sometimes a full chord of 100 ft.) to $P.T.$ The total distance from $P.C.$ to $P.T.$ measured in this way, is the **Length of Curve**, L.

39. The degree of curve is here defined as the angle subtended by a *chord* of 100 feet, rather than by an *arc* of 100 feet.

Either assumption involves the use of approximate methods either in calculations or measurements, if the convenient and customary methods are followed. It is believed that on the merits of the question, it is best to accept the definition given, and the practice in this country is largely in harmony with this definition.

Outside of the United States a curve is generally designated by its Radius, R. In the United States for railroad purposes, a curve is generally designated by its Degree, D.

40. Problem. *Given R.*
Required D.

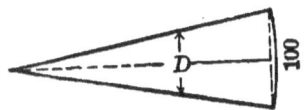

$$R \sin \tfrac{1}{2} D = \frac{100}{2}$$

$$\sin \tfrac{1}{2} D = \frac{50}{R} \quad (1)$$

41. Problem. *Given D.*
Required R.

$$R = \frac{50}{\sin \tfrac{1}{2} D} \quad (2)$$

Example. Given $D = 1°$.

$$R_1 = \frac{50}{\sin \tfrac{1}{2} D} \qquad \begin{array}{ll} 50 \log & 1.698970 \\ 0°\,30'\ \log \sin & 7.940842 \end{array}$$

$$R_1 = 5729.6 \qquad \log \quad 3.758128$$

42. Problem. *Given R_1 (radius of $1°$ curve) or D_1.*
Required R_a (radius of any given curve of degree $= D_a$).

$$R_1 = \frac{50}{\sin \tfrac{1}{2} D_1} \qquad R_a = \frac{50}{\sin \tfrac{1}{2} D_a}$$

$$\frac{R_a}{R_1} = \frac{\sin \tfrac{1}{2} D_1}{\sin \tfrac{1}{2} D_a} \qquad R_a = R_1 \frac{\sin \tfrac{1}{2} D_1}{\sin \tfrac{1}{2} D_a} \quad (3)$$

In the case of small angles, the angles are proportional to the sines (approximately),

$$R_a = R_1 \frac{\tfrac{1}{2} D_1}{\tfrac{1}{2} D_a} = R_a = R_1 \frac{D_1}{D_a}$$

But $R_1 = 5730$ to nearest foot,

$$R_a = \frac{5730}{D_a} \text{ (approx.)} \quad (4)$$

Example. $R_{10} = 573.7$ by (3), or by Searles' Table IV.
  $= 573.0$ by (4) (approx.)

Some engineers use shorter chords for sharp curves, as 1 to 7°, 100 ft.; 8° to 15°, 50 ft.; 16° to 20°, 25 ft. $R_a = \frac{5730}{D_a}$ is then very closely approximate.

Values of R and D are readily convertible. Table IV., Searles, serves this purpose, giving accurate results or values. In problems later, where either R or D is given, both will, in general, be assumed to be given. Approximate values can be found without tables by (4). The radius of a 1° curve = 5730 should be remembered.

43. Problem. *Given I, also R or D.*
Required T.

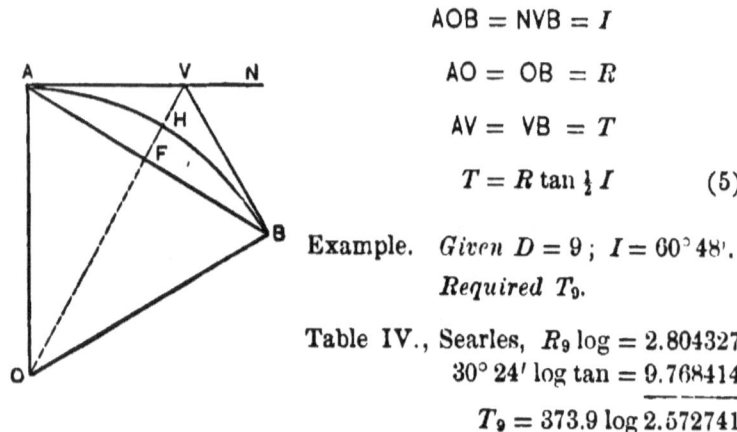

$AOB = NVB = I$

$AO = OB = R$

$AV = VB = T$

$$T = R \tan \tfrac{1}{2} I \qquad (5)$$

Example. *Given $D = 9$; $I = 60°48'$.*
Required T_9.

Table IV., Searles, $R_9 \log = 2.804327$
$30°24' \log \tan = 9.768414$
$T_9 = 373.9 \log 2.572741$

44. Approximate Method.

$$T_1 = R_1 \tan \tfrac{1}{2} I; \quad T_a = R_a \tan \tfrac{1}{2} I$$

$$\frac{T_a}{T_1} = \frac{R_a}{R_1} = \frac{D_1}{D_a} = \frac{1}{D_a} \text{ (approx.)}$$

$$T_a = \frac{T_1}{D_a} \text{ (approx.)} \qquad (6)$$

Table VI., Searles, gives values of T_1 for various values of I. ($I = \Delta$ of Searles).

Table V., Searles, gives a correction to be added *after* dividing by D_a.

Simple Curves.

Example. As before. Given $D = 9$; $I = 60° 48'$.
Required T_9.

Table VI., Searles,
$T_1\ 60° 40' = 3352.6$
$\underline{8' =\ \ \ \ 9.0}$
$T_1\ \overline{60° 48'} = 3361.6$
$T_9\ \ \ \ \ \ \ \ \ = 373.5$ (approx.)

Table V., Searles, correction, $\underline{.4}$
$T_9\ \ \ \ \ \ \ = 373.9$ (exact)
the same as before

(Interpolation for correction by inspection simply.)

45. Problem. *Given I, also R or D.*
Required E.

Using previous figure,

$$\mathsf{VH} = E = R \text{ exsec } \tfrac{1}{2} I \qquad (7)$$

Table XXIX., Searles, gives natural exsec.
Table XXVI., Searles, gives logarithmic exsec.
Approximate Method.
By method used for (6),

$$E_a = \frac{E_1}{D_a} \text{ (approx.)} \qquad (8)$$

Table VI., Searles, gives values for E_1.
Table V., Searles, gives correction to be used if desired.

46. Problem. *Given I, also R or D.*
Required M.

$$\mathsf{FH} = M = R \text{ vers } \tfrac{1}{2} I \qquad (9)$$

Table XXIX., Searles, gives natural vers.
Table XXVI., Searles, gives logarithmic vers.
Table VIII., Searles, gives certain middle ordinates.

47. Problem. *Given I, also R or D.*
Required chord $\mathsf{AB} = C$.

$$C = 2 R \sin \tfrac{1}{2} I \qquad (10)$$

Table VII., Searles, gives values for certain long chords.

48. Transposing, we find additional formulas, as follows:

from (5) $\quad R = T \cot \tfrac{1}{2} I \quad$ (11)

(7) $\quad R = \dfrac{E}{\operatorname{exsec} \tfrac{1}{2} I} \quad$ (12)

(9) $\quad R = \dfrac{M}{\operatorname{vers} \tfrac{1}{2} I} \quad$ (13)

(10) $\quad R = \dfrac{C}{2 \sin \tfrac{1}{2} I} \quad$ (14)

(4) $\quad D_a = \dfrac{5730}{R_a} \text{ (approx.)} \quad$ (15)

(6) $\quad D_a = \dfrac{T_1}{T_a} \text{ (approx.)} \quad$ (16)

(8) $\quad D_a = \dfrac{E_1}{E_a} \text{ (approx.)} \quad$ (17)

49. Problem. *Given sub-angle d, also R or D. Required sub-chord c.*

$$c = 2 R \sin \tfrac{1}{2} d \quad (18)$$

Approximate Method.

$$100 = 2 R \sin \tfrac{1}{2} D$$

$$\dfrac{c}{100} = \dfrac{\sin \tfrac{1}{2} d}{\sin \tfrac{1}{2} D} = \dfrac{d}{D} \text{ (approx.)} \quad (19)$$

The precise formula is seldom if ever used.

50. Problem. *Given sub-chord c, also R or D. Required sub-angle d.*

$$d = \dfrac{cD}{100} \quad (20)$$

$$\dfrac{d}{2} = \dfrac{c}{100} \dfrac{D}{2} \quad (21)$$

Simple Curves.

A modification of this formula is as follows:

$$\frac{d}{2} = \frac{cD}{200}$$

for $D = 1$

$$\frac{d}{2} = c\frac{60'}{200} = c \times 0.3'$$

for any value D_a

$$\frac{d}{2} = c \times 0.3' \times D_a \text{ (value in minutes)} \qquad (22)$$

This gives a very simple and rapid method of finding the value of $\frac{d}{2}$ in minutes, and the formula should be remembered.

Example. *Given sub-chord* $= 63.7$. $D = 6° 30'$.
Required sub-angle $d \left(or\ \frac{d}{2} \right)$.

I. By (20)   63.7
$\underline{\quad 6.5 = D}$
3185
3822
$\overline{414.05}$
4°.14
$\underline{\quad 60'}$
$d = 4° 08'$

$\frac{d}{2} = 2° 04'$

II. By (21)   63.7
$\underline{\quad 3.25 = \frac{D}{2}}$
3185
1274
1911
$\overline{207.025}$
2.07
$\underline{\quad 60'}$
$\frac{d}{2} = 2° 04'$

III. By (22)   63.7
$\underline{\quad 0.3}$
19.11
$\underline{\quad 6.5}$
9555
$\underline{11466\quad}$
$\overline{124.215}$ minutes

$\frac{d}{2} = 2° 04'$

Method III. seems preferable to I. or II.

51. Problem. *Given I and D.*
  Required L.

(a) When the *P.C.* is at an *even station*, *D* will be contained in *I* a certain number of times n, and there will remain a sub-angle d subtended by its chord c.

$$\frac{I}{D} = n + \frac{d}{D} = \quad n + \frac{c}{100} \text{ (approx.)}$$

$$100\frac{I}{D} = \quad = 100n + c = L \text{ (approx.)}$$

(b) When the *P.C.* is at a *sub-station* the same reasoning holds, and

$$L = 100\frac{I}{D} \text{ (approx.)} \tag{23}$$

Transposing,

$$I = \frac{LD}{100} \text{ (approx.)} \tag{24}$$

$$D = \frac{100\,I}{L} \tag{25}$$

These formulas (23)(24)(25), though approximate, are the formulas in common use.

Example. *Given 7° curve. $I = 39° 37'$.*
  Required L.

$$\begin{array}{rl} I = & 39° \; 37 \\ D = & 7\overline{)39.6167°} \\ & 5.6595+ \\ L = & 566.0 \end{array}$$

Example. *Given D and L.*
  Required I.
  Given 8° curve

$$\begin{array}{rl} \text{also, } P.T. = & 93 + 70.1 \\ P.C. = & 86 + 49.3 \\ L = & 7 \quad 20.8 \\ D = & \quad\;\; 8 \\ \hline & 57.664 \\ & \quad\;\; 60' \\ \hline & 39.84 \\ I = & 57° \; 40' \end{array}$$

Simple Curves.

52. Method of Deflection Angles.

If at any point on an existing curve a tangent to the curve be taken, the angles from the tangent to any given points on the curve may be measured, and the angles thus found may be called **Total Deflections** to those points (as NA1, NA2, NA3).

In laying out successive points upon a straight line (as on a "Tangent"), each point is generally fixed by

(a) measurement from the preceding point and

(b) line;

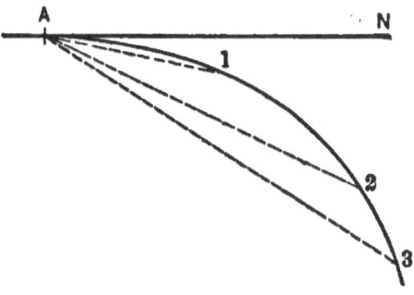

the line on a tangent will be the same for all points.

Similarly, in laying out a curve, successive points may be fixed by

(a) measurement from the preceding point and

(b) line;

the line in this case, for the curve, will be that found by using the *total deflection* calculated for each point. In the figure preceding, the chord distance A1 and the total deflection NA1 fix point 1; the chord distance 1-2 and total deflection NA2 fix point 2; and 2-3 and NA3 fix 3. A curve can be conveniently laid out by this method if the proper total deflections can be readily computed.

53. Problem. *Given a Parabolic Curve.*

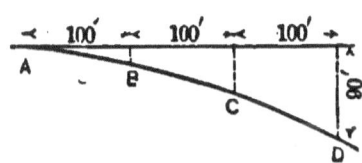

Required total deflections to B-C-D and chord lengths AB − BC − CD.

Give results to the nearest minute or nearest $\frac{1}{10}$ foot.

54. Simple Curves.

In the case of "*Simple Curves,*" the "total deflections" can be readily computed, and the method of "deflection angles" is therefore well adapted to laying them out.

55. Problem. To find the *Total Deflections* for a Simple Curve having *given* the *Degree*.

I. *When the curve begins and ends at even stations.*

The distance from station to station is 100 feet. The deflection angles are required.

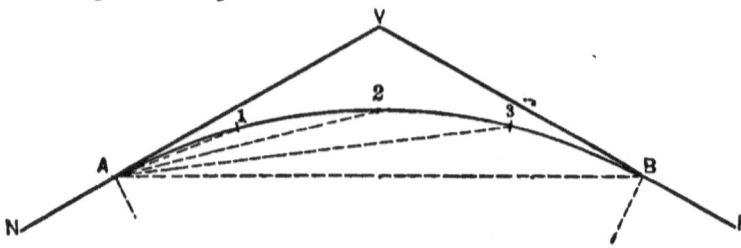

An acute angle between a tangent and a chord is equal to one half the central angle subtended by that chord

$$A\,1 = 100 \qquad V\,A\,1 = \tfrac{1}{2}D$$

The acute angle between two chords having their vertices in the circumference is equal to one half the arc included between those chords.

$$1 - 2 = 100 \quad \text{and} \quad 1\,A\,2 = \tfrac{1}{2}D \quad \text{Similarly,}$$
$$2 - 3 = 100 \quad \text{and} \quad 2\,A\,3 = \tfrac{1}{2}D$$
$$3 - B = 100 \quad \text{and} \quad 3\,A\,B = \tfrac{1}{2}D$$

This angle $\tfrac{1}{2}D$ is called by Henck and Searles the **Deflection Angle**, and will be so called here. Shunk and Trautwine call it the "*Tangential Angle*." The weight of engineering opinion appears to be largely in favor of the "Deflection Angle."

The "*Total Deflections*" will be as follows:

$$V\,A\,1 = \tfrac{1}{2}D$$
$$V\,A\,2 = V\,A\,1 + \tfrac{1}{2}D$$
$$V\,A\,3 = V\,A\,2 + \tfrac{1}{2}D$$

VAB will be found by successive increments of $\tfrac{1}{2}D$.

VAB = VBA = $\tfrac{1}{2}I$. This furnishes a "check" on the computation.

II. *When the curve begins and ends with a sub-chord.*

$$V\,A\,1 = \tfrac{1}{2}d$$
$$V\,A\,2 = V\,A\,1 + \tfrac{1}{2}D$$
$$V\,A\,3 = V\,A\,2 + \tfrac{1}{2}D$$

Simple Curves. 29

VAB is found by adding $\frac{1}{2} d_2$ to previous "total deflection."
VAB = VBA = $\frac{1}{2} I$. This furnishes "check." The total deflections should be calculated by successive increments; the final "check" upon $\frac{1}{2} I$ then checks all the intermediate total deflections. The example on next page will illustrate this.

56. Field-work of laying out a simple curve having given the position and station of *P.C.* and *P.T.*

(*a*) Set the transit at *P.C.* (A).
(*b*) Set the vernier at 0.
(*c*) Set cross hairs on V (or on N and reverse).
(*d*) Set off $\frac{1}{2} d_1$ (sometimes $\frac{1}{2} D$) for point l.
(*e*) Measure distance c_1 (sometimes 100) and fix l.
(*f*) Set off total deflection for point 2.
(*g*) Measure distance l–2 = 100 and fix 2, etc.
(*h*) When total deflection to B is figured, see that it = $\frac{1}{2} I$, thus "checking" calculations.
(*i*) See that the proper calculated distance c_2 and the total deflection to B agree with the actual measurements on the ground, checking the field-work.
(*k*) Move transit to *P.T.* (B).
(*l*) Turn vernier back to 0, and *beyond* 0 *to* $\frac{1}{2} I$.
(*m*) Sight on A.
(*n*) Turn vernier to 0.
(*o*) Sight towards V (or reverse and sight towards P), and see that the line checks on V or P.

It should be observed that three "checks" on the work are obtained.

1. The calculation of the total deflections is checked if total deflection to B = $\frac{1}{2} I$.

2. The chaining is checked if the final sub-chord measured on the ground = calculated distance.

3. The transit work is checked if the total deflection to B brings the line accurately on B.

The check in l is effective only when the total deflection for each point is found by adding the proper angle to that for the preceding point.

The check in 3 assures the general accuracy of the transit work, but does not prevent an error in laying off the total deflection at an intermediate point on the curve.

Railroad Curves and Earthwork.

57. Example. *Given Notes of Curve*

$$P.T.\ 13 + 45.0$$
$$P.C.\ 10 + 74.0 \quad 6° \text{ curve } L$$

Required the "total deflections"

$$\text{to sta. 11 } c_1 = \begin{array}{r} 26 \\ .3 \\ \hline 7.8 \end{array}$$

$$\frac{d_1}{2} = \frac{6°}{46.8} = 0°\ 47'\ \text{to 11}$$

$$c_2 = \begin{array}{r} 45 \\ .3 \\ \hline 13.5 \end{array} \quad \begin{array}{r} 3° \\ 3°\ 47'\ \text{to 12} \\ 3° \\ \hline 6°\ 47'\ \text{to 13} \end{array}$$

$$\frac{d_2}{2} = \frac{6°}{81.0} = \begin{array}{r} 1°\ 21' \\ \hline 8°\ 08'\ \text{to } 13 + 45 \end{array}$$

$$\begin{array}{r} 13 + 45.0 \\ 10 + 74.0 \\ \hline 2 \quad 71.0 = L \end{array}$$

$$\begin{array}{r} 6° \\ \hline 16.26 \\ 60' \\ \hline 15.6' \end{array} \quad 16°$$

$$\begin{array}{r} 16' \\ \hline 16°\ 16' = I \\ 8°\ 08' = \tfrac{1}{2} I \text{ "check"} \end{array}$$

58. Caution.

If a curve of nearly $180° = I$ is to be laid out from A, it is evident that it would be difficult or impossible to set the last point accurately, as the "intersection" would be bad. It is undesirable to use a total deflection greater than 30°.

It may be impossible to see the entire curve from the *P.C.* at A.

It will, therefore, frequently happen that from one cause or another the entire curve cannot be laid out from the *P.C.*, and it will be necessary to use a modification of the method described above.

Simple Curves.

59. Field-work. *When the entire curve cannot be laid out from the P.C.*

First Method.

(*a*) Lay out curve as far as C, as before.

(*b*) Set transit point at some convenient point, as C (even station preferably).

(*c*) Move transit to C.

(*d*) Turn vernier back to 0, *and beyond* 0 to measure the angle VAC.

(*e*) Sight on A.

(*f*) Turn vernier to 0. See that transit line is on auxiliary tangent NCM (VAC = NCA being measured by ½ arc AC).

(*g*) Set off new deflection angle (½ *d* or ½ *D*).

(*h*) Set point 4, and proceed as in ordinary cases.

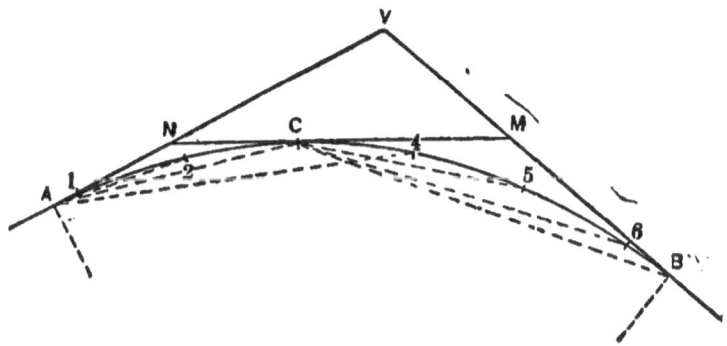

60. Second Method.

(*a*) Set point C as before, and move transit to C.

(*b*) Set vernier at 0.

(*c*) Sight on A.

(*d*) Set off the proper "total deflection" for the point 4 = VA 4. NCA + MC 4 = VA 4, each measured by ½ arc AC 4.

(*e*) Reverse transit and set point 4.

(*f*) Set off and use the proper "total deflections" for the remaining points.

The second method is in some respects more simple, as the notes and calculations, and also setting off angles, are the same as if no additional setting were made. By the first method the deflection angles to be laid off will, in general, be even minutes, often degrees or half degrees, and are thus easier to lay off. It is a matter of personal choice which of the two methods shall be used. It will be disastrous to attempt an incorrect combination of parts of the two methods.

61. Field-work of finding $P.C.$ and $P.T.$

In running in the line, it is common and considered advisable to establish "V," determine the station of "V," and measure the angle I. Having given I only, an infinite number of curves could be used. It is, therefore, necessary to assume additional data to determine what curve to use. It is common to proceed as follows:

(a) Assume either (1) D directly.

(2) E and calculate D.

(3) T and calculate D.

It is often difficult to determine off-hand what degree of curve will well fit the ground. Frequently the value of E_a can be readily determined on the ground. The determination of D from E_a is readily made, using the approximate formula $D_a = \dfrac{E_1}{E_a}$. Similarly, we may be limited to a given (or ascertainable) value of T_a, and from this readily find $D_a = \dfrac{T_1}{T_a}$.

The value of D_a adopted will, in general, be taken to the nearest $\frac{1}{2}°$ (perhaps only to nearest degree) rather than at the exact value found, as above. (Some engineers use $1° 40' = 100'$ and $3° 20' = 200'$, etc., rather than $1° 30'$ or $3° 30'$, etc.).

(b) From the data finally adopted T is *calculated anew*.

(c) The instrument still being at V, the $P.T.$ is set by laying off T.

(d) The station of $P.C.$ is calculated and $P.C.$ set.

(e) The length of curve L is calculated, and station of $P.T.$ thus determined (not by adding T to station of V).

Total deflections should be all calculated and entered in notebook.

Whether D, E, or T shall be assumed depends upon the special requirements in each case. Curves are often run out from $P.C.$ without finding or using V, but the best engineering usage seems to be in favor of setting V, whenever this is at all practicable, and from this finding the $P.C.$ and $P.T.$

Simple Curves. 33

62. Example. *Given a line, as shown in sketch.*
Required a Simple Curve to connect the Tangents.

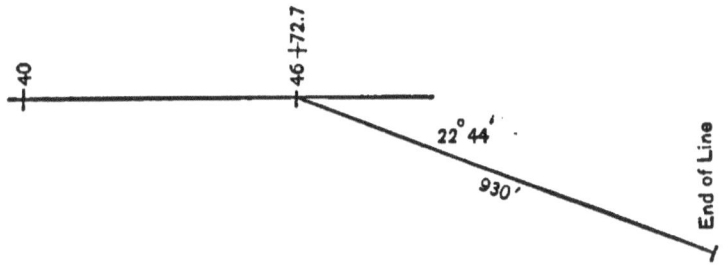

P.T. is to be at least 400 ft. from end of line.
Use smallest degree or half degree consistent with this.
Find degree of curve and stations of P.C. and P.T.

Table VI, Searles, $22° 40'\ T_1 = 1148.4$ 930
                              4'   3.4   400
                           ──────   ──────   ────────
                           22° 44'   1151.8   (530 = T_a
                                        1060   (2.2 −
                                        ──────
                                         918

                                              use 2° 30' curve
                            $T_1 = 1151.8$ (2.5°
                                    100   (460.7
                                   ──────
                                    151
                                    150
                                   ──────
                                    180
                    Table V, correction   0
                                             ──────
                                             460.7 = T.

$V = 46 + 72.7$
$T \quad 4 + 60.7$ $22° 44' = I$
$P.C. \quad 42 + 12.0$ $2.5°) 22° 7333$
$L \quad 9 + 09.3$ $909.3 = L$
$P.T. \quad 51 + 21.3$

63. Form of Transit Book (left-hand page).

(Date)
(Names of Party)

Station	Points	Descrip. of Curve	Total Deflect.	Observed Course
114				
113				
112				
111				
110				
109	⊙+ 90.0 *P.T.*		11° 15′	N 46° 00″ E
108		$R = 1146.3$	9° 00′	
107		$L = 450.0$	6° 30′	
106	⊙+ 68.0 *V*	$T = 228.0$	4° 00′	
105		$I = 22° 30′$	1° 30′	
104	⊙+ 40.0 *P.C.*	5° Right		
103				
102				
101				
100				
99				N 23° 15′ E
98				

V is not a point on the curve. Nevertheless, it is customary to record the station found by chaining along the tangent.

The right-hand page is used for survey notes of crossings of fences and various similar data. It seems unnecessary to show a sample here.

Simple Curves. 35

64. Metric Curves.

In Railroad Location under the "**Metric System**" a chain of 100 meters is too long, and a chain of 10 meters is too short. Some engineers have used the 30-meter chain, some the 25-meter chain, but lately the 20-meter chain has been generally adopted as the most satisfactory. Under this system a "*Station*" is 10 meters. Ordinarily, every second station only is set, and these are marked Sta. 0, Sta. 2, Sta. 4, etc. On curves, *chords of 20 meters* are used. Usage among engineers varies as to what is meant by the *Degree of Curve* under the metric system. There are two distinct systems used, as shown below.

I. The *Degree of Curve* is the *angle at the center* subtended by a chord of 1 chain of *20 meters*.

II. The *Degree of Curve* is the *deflection angle* for a chord of 1 chain of *20 meters* (or one half the angle at the center).

II. Or, very closely, the *Degree of Curve* is the *angle at the center* subtended by a chord of *10 meters* (equal to 1 station length).

For several reasons the latter system is favored here. Tables upon this basis have been calculated, giving certain data for metric curves. Such tables are to be found in Henck's Field-Book, but not in Searles.

In many countries where the metric system is used, it is not customary to use the *Degree of Curve*, as indicated here. In Mexico, where the metric system is adopted as the only legal standard, very many of the railroads have been built by companies incorporated in this country, and under the direction of engineers trained here. The usage indicated above has been the result of these conditions. If the metric system shall in the future become the only legal system in the United States, as now seems possible, one of the systems outlined above will probably prevail.

In foreign countries where the *Degree of Curve* is not used, it is customary, as previously stated, to designate the curve by its radius *R*, and to use even figures, as a radius of 1000 feet, or 2000 feet, or 1000 meters, or 2000 meters. As the radius is seldom measured on the ground, the only convenience in even figures is in platting, while there is a constantly recurring inconvenience in laying off the angles.

65. Problem. *Given D and the stations of P.C. and P.T. Required to lay out the curve by the method of* **Deflection Distances.**

I. *When the curve begins and ends at even stations.*

In the curve AB, let

AN be a tangent

AE any chord $= c$

EE′ perp. to AE′ $= a =$

"tangent deflection"

FF′ $=$ BB′ $=$ the

"chord deflection"

AO $=$ EO $= R$

Draw OM perpendicular to AE.

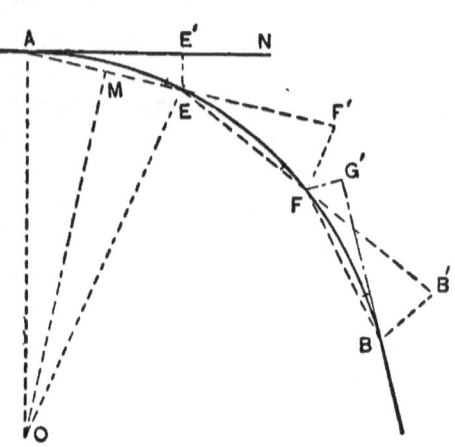

Then
$$EE' : AE = ME : EO$$
$$a : c = \frac{c}{2} : R$$
$$a = \frac{c^2}{2R} \qquad (26)$$

FF′ $= 2a$; AF′ $=$ AE produced

When AE is a full station of 100 feet,
$$a_{100} = \frac{100^2}{2R}$$

66. Field-work.

The *P.C.* and *P.T.* are assumed to have been set.

(*a*) Calculate a_{100}.

(*b*) Set point E distant 100 ft. from A and distant a_{100} from AE′ (AE′ $<$ 100 ft. ; AE′E $= 90°$).

(*c*) Produce AE to F′ (EF′ $=$ 100 ft.), and find F distant $2 a_{100}$ from F′ (EF $=$ 100 ft.).

(*d*) Proceed similarly until B is reached (*P.T.*).

Simple Curves. 37

(e) At station preceding B (P.T.) lay off $FG' = a_{100}$ (FG'B = 90°).

(f) G'B is tangent to the curve at B (P.T.).

67. The tangent deflection for 100 ft. for a 1° curve is nearly 0.875, or $\frac{7}{8}$ ft. For any number of even stations n, the offset will be

$$a_n = \tfrac{7}{8} n^2 \text{ for a 1° curve (approx.).}$$
$$a_n = \tfrac{7}{8} n^2 D_a \text{ for any curve (approx.)} \qquad (27)$$

68. Problem. *Given two Curves of degree D_i and D_f. Required the offset between the two Curves for any number of even stations.*

From (27)
$$a_{ni} = \tfrac{7}{8} n^2 D_i$$
$$a_{nf} = \tfrac{7}{8} n^2 D_f$$
$$a_n = \tfrac{7}{8} n^2 (D_f - D_i) \text{ (approx.)} \qquad (28)$$

69. Problem. *Given D and the stations of P.C. and P.T. Required to lay out the Curve by Deflection Distances.*

II. *When the curve begins and ends with a sub-chord.*

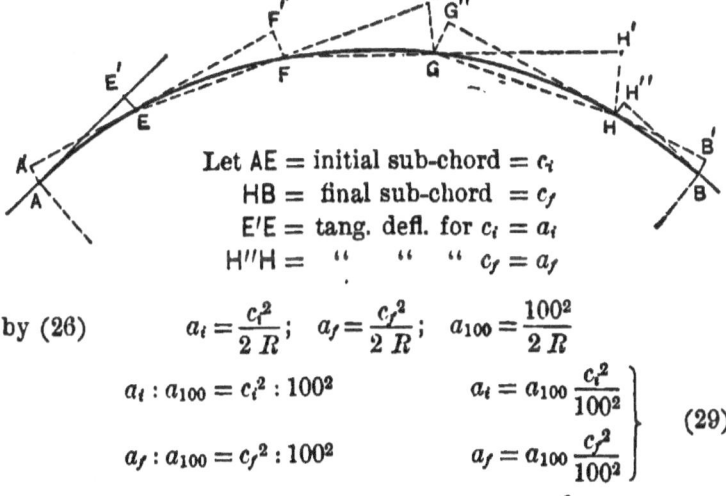

Let AE = initial sub-chord = c_i
HB = final sub-chord = c_f
E'E = tang. defl. for $c_i = a_i$
H''H = " " " $c_f = a_f$

by (26) $\quad a_i = \dfrac{c_i^2}{2R}; \quad a_f = \dfrac{c_f^2}{2R}; \quad a_{100} = \dfrac{100^2}{2R}$

$$\left.\begin{array}{l} a_i : a_{100} = c_i^2 : 100^2 \\[4pt] a_f : a_{100} = c_f^2 : 100^2 \end{array}\quad \begin{array}{l} a_i = a_{100} \dfrac{c_i^2}{100^2} \\[4pt] a_f = a_{100} \dfrac{c_f^2}{100^2} \end{array}\right\} \qquad (29)$$

In general it is better to use (29) than $a_i = \dfrac{c_i^2}{2R}$.

70. Example. Given P.T. $20 + 42$ $\quad$ 6° curve R
$\qquad\qquad\qquad$ P.C. $16 + 25$

Required all data necessary to lay out curve by "*Deflection Distances.*"

Calculate without Tables. Result to $\frac{1}{100}$ foot.

Radius 1° curve $= \dfrac{5730(6}{955}$
$\qquad\qquad\quad 6°$

$a_{100} = \dfrac{100^2}{2 \times 955}$

$\qquad = 5.24$

$2\, a_{100} = 10.47$

$a_{75} = 0.75^2 \times 5.24$

$\qquad = 2.95$

$a_{42} = 0.42^2 \times 5.24$

$\qquad = 0.92$

$\begin{array}{r}1910\,)\,10000\,(5.235+\\\underline{955}\\450\\\underline{382}\\680\\\underline{573}\\1070\\\underline{955}\end{array}$

Searles' Tables give $a_{100} = 5.234$ (precise value).

71. The distance AE′ is slightly shorter than AE. It is generally sufficient to take the point E′ by inspection simply. If desired for this or any other purpose, a simple approximate solution of right triangles is as follows:

Problem. Given the hypoteneuse (or base) and altitude.
$\qquad\quad$ Required the difference between base and hypoteneuse, or in the figure, $c - a$.

$c^2 - a^2 \qquad\quad = h^2$

$(c - a)(c + a) = h^2$

$$c - a = \dfrac{h^2}{c + a} = \dfrac{h^2}{2c} \text{ (approx.)}$$

$$c - a = \dfrac{h^2}{2a} \text{ (approx.)} \qquad (30)$$

Wherever h is small in comparison with a or c, the approximation is good for ordinary purposes.

Example. $\qquad c = 100 \quad h = 10$

$\qquad\qquad\quad c - a = \tfrac{100}{200} = \;\; 0.50$

$\qquad\qquad\quad a \quad = \qquad\quad 99.50$

The precise formula gives 99.499.

Simple Curves. 39

72. Field-work *for Case II., p. 37.*

(*a*) Calculate a_{100}, a_i, a_f. Remember that tangent deflections are as the *squares* of the chords.

a_{100} may be found generally in Table IV., Searles, as "tangent offset."

(*b*) Find the point E, distant a_i from AE' and distant c_i from A. (AE'E = 90°.)

(*c*) Erect auxiliary tangent at E (lay off AA' = a_i).

(*d*) From auxiliary tangent A'E produced, find point F.

(FF' = a_{100}; EF = 100; EF'F = 90°).

(*e*) From chord EF produced, find point G.

(GG' = 2 a_{100}; FG' = FG = 100).

(*f*) Similarly, for each full station, use 2 a_{100}, etc.

(*g*) At last even station on curve, H, erect an auxiliary tangent (lay off GG'' = a_{100}; GG''H = 90°).

(*h*) From G''H produced, find B (B'B = a_f, etc.).

(*i*) Find tangent at B (HH'' = a_f; HH''B = 90°).

The values of a_{100}, a_i, a_f, should be calculated to the nearest $\frac{1}{100}$ foot.

73. Caution. The tangent deflections vary as the *squares* of the chords, not directly as the chords.

Curves may be laid out by this method without a transit by the use of plumb line or "flag" for sighting in points, and with *fair* degree of accuracy.

For calculating a_{100}, a_i, a_f, it is sufficient in most cases to use the approx. value $R_a = \dfrac{5730}{D_a}$. A curve may be thus laid out without the use of transit or tables.

For many approximate purposes it is well and useful to remember that the "chord deflection" for 1° curve is 1.75 ft. nearly, and for other degrees in direct proportion. A head chainman may thus put himself *nearly* in line without the aid of the transitman.

The method of "Deflection Distances" is not well adapted for common use, but will often be of value in emergencies.

74. Problem. *Given D* and stations of *P.C.* and *P.T.*
Required to lay out the curve by "*Deflection Distances*" when the *first sub-chord is small.*

Caution. It will not be satisfactory in this case to produce the curve from this short chord.

75. Problem. *Given D* and stations of *P.C.* and *P.T.*
Required to lay out the curve by the method of **Offsets from the Tangent.**

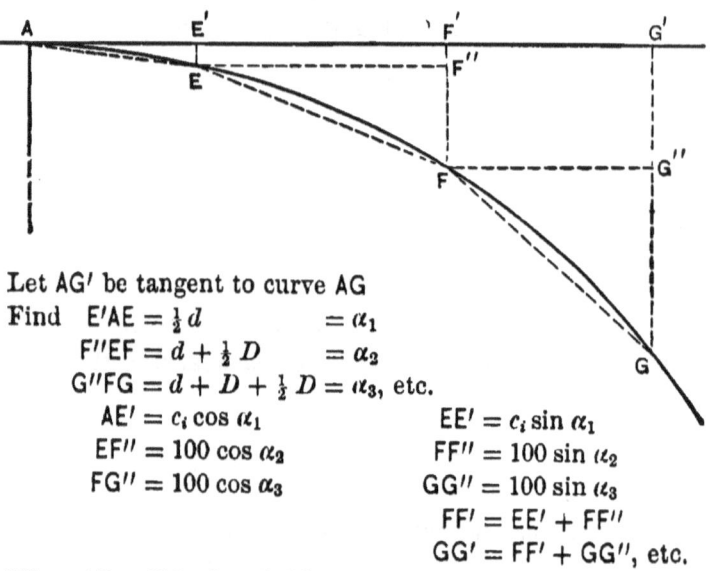

Let AG′ be tangent to curve AG
Find E′AE = $\tfrac{1}{2} d$ = α_1
 F″EF = $d + \tfrac{1}{2} D$ = α_2
 G″FG = $d + D + \tfrac{1}{2} D = \alpha_3$, etc.

AE′ = $c_i \cos \alpha_1$ EE′ = $c_i \sin \alpha_1$
EF″ = $100 \cos \alpha_2$ FF″ = $100 \sin \alpha_2$
FG″ = $100 \cos \alpha_3$ GG″ = $100 \sin \alpha_3$
     FF′ = EE′ + FF″
     GG′ = FF′ + GG″, etc.

When AE = 100, then $\tfrac{1}{2} d$ becomes $\tfrac{1}{2} D$.

76. Field-work.

(*a*) Calculate   AE′, E′F′, F′G′
              EE′, FF′, GG′

(*b*) Set E′, F′, G′.

(*c*) Set E by distance AE (c_i) and EE′.

(*d*) Set F " " EF (100) and FF′.

(*e*) Set G " " FG (100) and GG′.

For the computations indicated above, always use natural sines and cosines.

Simple Curves.

77. Ordinates.

Problem. *Given D and two points on a curve.*

Required the *Middle Ordinate* from the chord between those two points.

By (9), $M = R \text{ vers } \tfrac{1}{2} I$

for 100 ft. chord $M = R \text{ vers } \tfrac{1}{2} D$

between points 200 ft. apart $M = R \text{ vers } D.$

Let $A =$ angle at center between the two points.

$$M = R \text{ vers } \tfrac{1}{2} A.$$

Table VIII., Searles, gives middle ordinates for certain long chords.

78. Problem. *Given R and C.*
Required M.

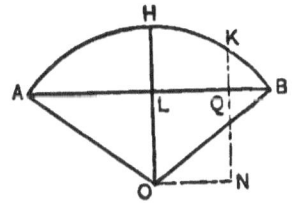

$$\text{OL} = \sqrt{R^2 - \left(\frac{c}{2}\right)^2}$$

$$\text{HL} = M = R - \sqrt{R^2 - \left(\frac{c}{2}\right)^2} \quad (31)$$

$$M = R - \sqrt{\left(R - \frac{c}{2}\right)\left(R + \frac{c}{2}\right)} \quad (32)$$

Table XXIII., Searles, gives squares and square roots for certain numbers. If the numbers to be squared can be found in this table, use (31). Otherwise use logarithms and (32).

79. Problem. *Given R and C.*
Required the *Ordinate* at any given point Q.

Measure $\text{LQ} = q.$ Then $\text{KN} = \sqrt{R^2 - q^2}$

$$\text{LO} = \sqrt{R^2 - \left(\frac{c}{2}\right)^2}$$

$$\text{KQ} = \text{KN} - \text{LO} = \sqrt{(R+q)(R-q)} - \sqrt{\left(R + \frac{c}{2}\right)\left(R - \frac{c}{2}\right)} \quad (33)$$

80. When $C = 100$ ft. or less, an approximate formula will generally suffice.

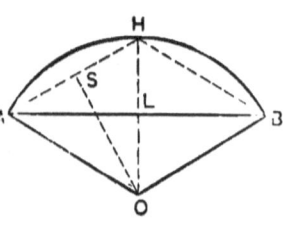

Problem. *Given R and c.*
Required M (approx.)

$$HL = M : AH = \frac{AH}{2} : R$$

$$M = \frac{AH^2}{2R}.$$

Where AB is small compared with R,

$$AH = \frac{c}{2} \text{ (approx.)}$$

$$M = \frac{c^2}{8R} \text{ (approx.)} \qquad (34)$$

81. Example. *Given $C = 100$, $D = 9°$.*
Required M.

$$R_9 = \frac{5730}{9} = 636.7$$

$$\phantom{R_9 = \frac{5730}{9} =} 8$$

$$5093.6)\,10000.\,(1.963 = M$$

Precise value
$M = 1.965$

$$\begin{array}{r} 50936 \\ \hline 490640 \\ 458424 \\ \hline 332160 \\ 305616 \\ \hline 26544 \end{array}$$

Table XII., Searles, gives middle ordinates for curving rails of certain lengths.

82. Problem. *Given R and c.*
Required Ordinate at any given point Q.
Approximate Method.

I. Measure $LQ = q$

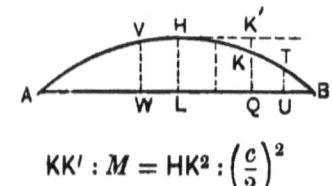

$$M = HL = \frac{\left(\frac{c}{2}\right)^2}{2R}$$

$$KK' = \frac{HK^2}{2R}$$

$$KK' : M = HK^2 : \left(\frac{c}{2}\right)^2$$

Simple Curves.

Since HK $= q$ (approx.) $KK' = \dfrac{q^2}{\left(\dfrac{c}{2}\right)^2} M$ (approx.) (35)

$$KQ = M - KK'$$

When $\dfrac{q}{\dfrac{c}{2}} = \dfrac{1}{2}$ as in figure, $KK' = \dfrac{M}{4}$ and $KQ = \dfrac{3}{4} M$ (approx.)

When $\dfrac{q}{\dfrac{c}{2}} = \dfrac{1}{4}$ $VW = \dfrac{15}{16} M$ (approx.)

When $\dfrac{q}{\dfrac{c}{2}} = \dfrac{3}{4}$ $TU = \dfrac{7}{16} M$ (approx.)

The curve thus found is accurately a parabola, but for short distances this practically coincides with a circle.

83. II. *Approximate Method.* Measure LQ and QB

$$M = \dfrac{\left(\dfrac{c}{2}\right)^2}{2R} \qquad KK' = \dfrac{q^2}{2R} \text{ (approx.)}$$

$$KQ = \dfrac{\left(\dfrac{c}{2}\right)^2 - q^2}{2R} = \dfrac{\left(\dfrac{c}{2}+q\right)\left(\dfrac{c}{2}-q\right)}{2R} \text{ (approx.)}$$

$$KQ = \dfrac{AQ \times QB}{2R} \text{ (approx.)} \qquad (36)$$

Sometimes one, sometimes the other of these methods will be preferable.

84. Example. *Given* $C = 100$, $D = 9°$.

$M = 1.965$ from Tables.

Required, Ordinate at point 30 ft. distant from center toward end of chord.

I. 30 ft. $= \dfrac{30}{50} \times \dfrac{c}{2}$

$KK = \tfrac{9}{25} \times 1.965$

$$\begin{array}{r} 9 \\ \hline 25)\,\overline{17.685} \\ .70740 \end{array}$$

$M = 1.965$
Ordinate $= 1.258$
Precise result for data above $= 1.260$.

II. $AQ = 80$
$BQ = 20$
 $1273.4)\,\overline{1600}(1.257$

$R_1 = 5730.$
$R_9 = 636.7$
$2R_9 = 1273.4$

$$\begin{array}{r} 1273.4 \\ \hline 32660 \\ 25468 \\ \hline 71920 \\ 63670 \\ \hline 8250 \end{array}$$

85. Problem. *Given R and c.*

Required a series of points on the curve.

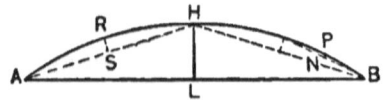

$$M = \mathsf{HL} = \frac{c^2}{8R} \text{ (approx.)}$$

$$\mathsf{RS} = \frac{\mathsf{AH}^2}{8R} \text{ (approx.)} \qquad \mathsf{AH} = \frac{c}{2} \text{ (approx.)}$$

$$\mathsf{RS} = \frac{\frac{c^2}{4}}{8R} = \frac{M}{4} \text{ (approx.)}$$

$$\mathsf{PN} = \frac{\mathsf{RS}}{4} \text{ (approx.), etc., as far as desirable.}$$

This method is useful for many general purposes, for ordinates in bending rails among others.

86. Problem. *Given a Simple Curve joining two tangents.*
Required the *P.C.* of a *new curve* of the *same radius* which shall end in a *parallel tangent.*

Let AB be the given curve.

A'B' " " required curve.

B'E = p = perpendicular distance between tangents.

Join BB'.

Then AA' = OO' = BB'

Also B'BE = V'VB = I

BB' sin I = p

$$\mathsf{BB'} = \mathsf{AA'} = \frac{p}{\sin I} \qquad (37)$$

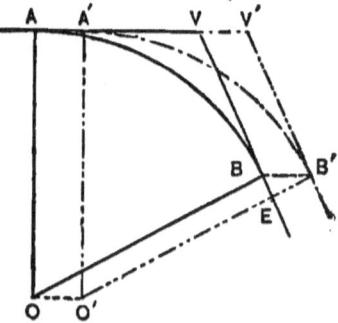

When the proposed tangent is *outside* the original tangent, the distance AA is to be added to the station of the *P.C.* When *inside*, it is to be subtracted.

Simple Curves. 45

87. Problem. *Given a Simple Curve joining two tangents. Required the Radius of a new curve which with the same P.C. shall end in a parallel tangent.*

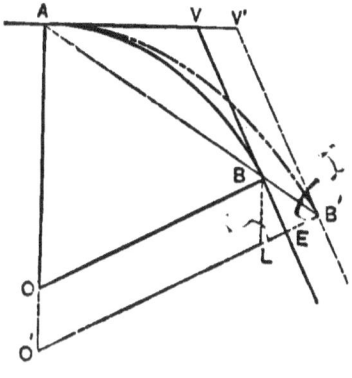

Let AB be the given curve of radius $R = AO$.

$B'E = p =$ perpendicular distance.

AB' the required curve, radius $= R'$.

Draw chords AB, AB';

also line BB';

also BL parallel to AO'.

Then
$$BL = OO'$$
$$= R' - R$$
$$= B'L$$
$$BLB' = AO'B'$$
$$= I$$
$$BL \text{ vers } BLB' = B'E$$
$$(R' - R) \text{ vers } I = p$$
$$(R' - R) = \frac{p}{\text{vers } I} \tag{38}$$

Since $VAB = V'AB'$, AB and AB' are in the same straight line.
$$BB' = 2 BL \sin\tfrac{1}{2} BLB'$$
$$= 2(R' - R) \sin\tfrac{1}{2} I \tag{39}$$

When the proposed tangent is *outside* the original tangent (as it is shown in the figure), the above formula applies, and
$$R' > R.$$

When the proposed tangent is *inside* the original tangent, the formula becomes
$$R - R' = \frac{p}{\text{vers } I} \tag{40}$$
and $R' < R$.

46 Railroad Curves and Earthwork.

88. Problem. *Given a Simple Curve joining two tangents. Required the radius and P.C. of a new curve to end in a parallel tangent with the new P.T. directly opposite the old P.T.*

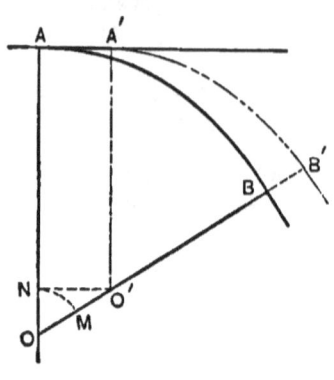

Let AB be the given curve of radius = R.

A'B' the required curve of radius R'.

BB' = p.

Draw perpendicular O'N and arc NM

Then O'M = B'M − B'O'
     = B'M − BM = BB'

O'M = p

ON exsec NOO' = O'M

$(R − R')$ exsec I = p; $R − R' = \dfrac{p}{\text{exsec } I}$ (41)

AA' = O'N = ON tan NOO'

AA' = $(R − R')$ tan I (42)

When the new tangent is *outside* the original tangent (as in the figure), $R > R'$ and AA' is added to the station of the P.C.

When the new tangent is *inside* the original tangent, $R < R'$, $R' − R = \dfrac{p}{\text{exsec } I}$, and AA' is subtracted from station of P.C.

89. Problem. *To find the Simple Curve that shall join two given tangents and pass through a given point.*

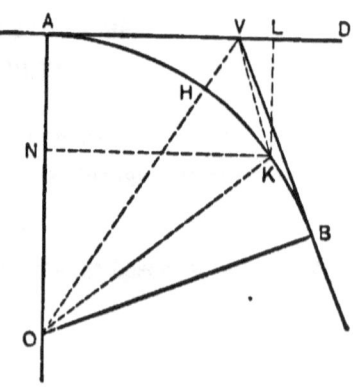

With the transit at V, the given point K can often be best fixed by angle BVK and distance VK. If the point K be fixed by other measurements, these generally can readily be reduced to the angle BVK and distance VK.

Simple Curves.

90. Problem. *Given the two tangents intersecting at V, the angle I, and the point K fixed by angle BVK = β and distance VK = b.*
Required the radius R of curve to join the two tangents and pass through K.

In the triangle VOK we have given

$$VK = b \text{ and } OVK = \frac{180 - I}{2} - \beta$$

Further $VO = \dfrac{R}{\cos \frac{1}{2} I}$ $OK = R$

$VO : OK = \sin VKO : \sin OVK$

$\dfrac{R}{\cos \frac{1}{2} I} : R = \sin VKO : \cos(\frac{1}{2} I + \beta)$

$$\sin VKO = \frac{\cos(\frac{1}{2} I + \beta)}{\cos \frac{1}{2} I} \tag{43}$$

From data thus found, the triangle VOK may be solved for R.
In solving this triangle the angle VOK is often very small. A slight error in the value of this small angle may occasion a large error in the value of R. In this case use the following **Second Method** of finding R after VOK has been found.

Find AOK $= \frac{1}{2} I + $ VOK   Also DVK $= I + \beta$
Then R vers AOK $=$ LK
$\qquad\qquad\qquad = b \sin $ DVK

$$R = \frac{b \sin \text{DVK}}{\text{vers AOK}} \tag{44}$$

91. Problem. *Given R, I, β (BVK).*
Required b (VK).

In the triangle VOK

$$OK = R; \quad OV = \frac{R}{\cos \frac{1}{2} I}$$

$$OVK = 90 - (\tfrac{1}{2} I + \beta)$$

Solve triangle for b.
Also find VOK and station of K if desired.

48   *Railroad Curves and Earthwork.*

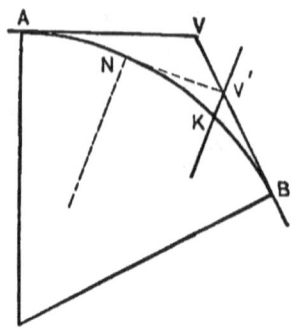

92. Problem. *To find the point where a straight line intersects a curve between stations.*

Find where the straight line V'K cuts VB at V'.

Measure KV'B.

Use V' as an auxiliary vertex.

Find I' from V'B.

Solve by preceding problem.

93. Approximate Method.

Set the middle point H by method of ordinates.

If the arc HB is sensibly a straight line, find the intersection of HB and CD.

Otherwise set the point G by method of ordinates, and get intersection of HG and CD.

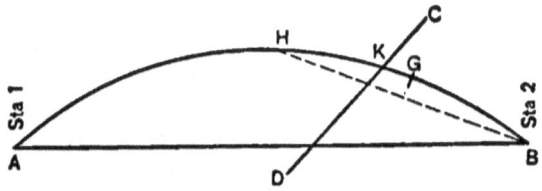

Additional points on the arc may be set if necessary, and the process continued until the required precision is secured.

The points H and G can be set without the use of a transit with sufficient accuracy for many purposes, a plumb line or flag being used in "sighting in."

94. Problem. *Given a Simple Curve and a point outside the curve.*

Required a tangent to the curve from that point.

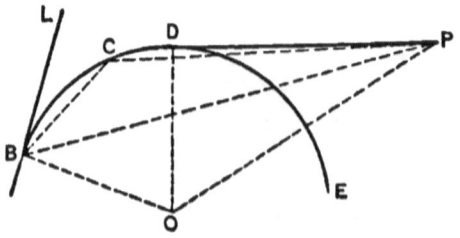

Let BDE be the given curve.

P the point outside the curve.

BL a tangent at B.

Measure LBP, also BP.

Simple Curves.

In the triangle BPO we have given PBO, BP, BO.

Solve the triangle for BOP and OP.

Then $$\cos DOP = \frac{OD}{OP} = \frac{R}{OP}$$

$$BOD = BOP - DOP$$

From BOD find station of D from known point B.

It should be noted that if log OP is found, this can be used again without looking out the number for OP. Other similar cases will occur elsewhere in calculation.

When for any reason it is difficult or inconvenient to measure BP directly, the angles CBP, BCP and the distance BC may be measured and BP calculated.

95. Approximate Method.

Field-work.

(a) From the station (B) nearest to the required point D, find by the approximate method where BP cuts the curve at C. (If C be the nearest station, produce PC to B.)

(b) Assume D with BD slightly greater than CD, and with transit at $P.C.$ set the point D (transit point) truly on the curve.

(c) Move the transit to D, and lay off a tangent to the curve at D. This will very nearly strike P.

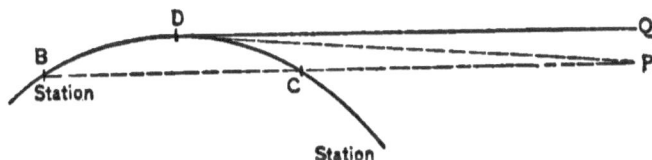

(d) If the tangent strikes away from P, at Q, measure QDP, and move the point D (ahead or back as the case may be) a distance c due to an angle at the center $d =$ QDP. The tangent from this new point ought to strike P almost exactly.

In a large number of cases the point D will be found on the first attempt sufficiently close for the required purpose.

If a tangent between two curves is required, similar methods by approximation will be found available.

Obstacles.

When any obstacle occurs upon a "tangent," the ordinary methods of surveying for passing such obstacle will be used.

96. When *V* is inaccessible.

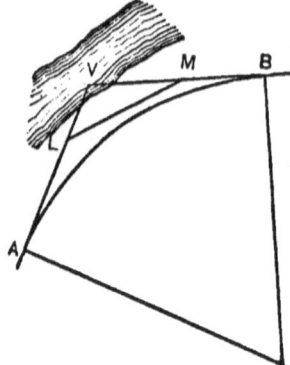

Measure VLM, VML, LM.

$I = $ VLM $+$ VML

LV and VM are readily calculated, and AL and MB determined.

In some cases the best way is to assume the position of *P.C.* and run out the curve as a trial line, and finally find the position of *P.C.* correctly by the method of formula (37).

97. When the *P.C.* is inaccessible.

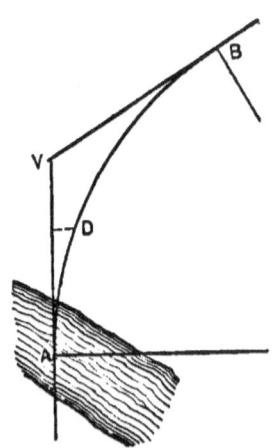

Establish some point D (an even station is preferable) by method of "offsets from Tangent" or otherwise.

Move transit to B (*P.T.*), and run out curve starting from D and checking on tangent VB.

98. When the *P.T.* is inaccessible.

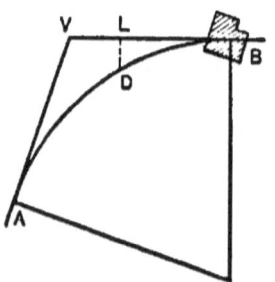

With instrument still at *V*, set some convenient point D, move transit to *P.C.*, and run in curve to D, and then pass the obstacle at B as any obstacle on a tangent would be passed.

Simple Curves.

99. When **Obstacles on the Curve** occur so as to prevent running in the curve, no general rules can well be given. Sometimes *resetting* the transit in the curve will serve. Sometimes, if one or two points only are invisible from the transit, these can be set by "*deflection distances*," and the curve continued by "*deflection angles*," without resetting the transit. Sometimes "*offsets from the tangent*" can be used to advantage. Sometimes points can be set by "*ordinates*" from chords. Sometimes the method shown on page 48, § 92, assuming an auxiliary V, is the only one possible.

It should be borne in mind that it is seldom *necessary* that the *even stations* should be set. If it be possible to set any points whose stations are known and which are not too far apart, this is generally sufficient.

Finally, for passing obstacles and for solving many problems which occasionally occur, it is necessary to understand the various methods of laying out curves, and to be familiar with the mathematics of curves; and, in addition, to exercise a reasonable amount of ingenuity in the application of the knowledge possessed.

CHAPTER V.

COMPOUND CURVES.

100. When one curve joins another, the two curves having a common tangent at the point of junction, and lying upon the same side of the common tangent, the two curves form a *Compound Curve*.

When two such curves lie upon opposite sides of the common tangent, the two curves then form a *Reversed Curve*.

In a compound curve, the point at the common tangent where the two curves join, is called the $P.C.C.$, meaning the "point of compound curvature."

In a reversed curve, the point where the curves join is called the $P.R.C.$, meaning the "point of reversed curvature."

101. Field-work.

Laying out a compound curve or a reversed curve.

(*a*) Set up transit at $P.C.$

(*b*) Run in simple curve to $P.C.C.$ or $P.R.C.$

(*c*) Move transit to $P.C.C.$ or $P.R.C.$

(*d*) Set line of sight on common tangent with vernier at 0.

(*e*) Run out second curve as a simple curve.

It is not desirable to attempt to lay out the second curve with the transit at the $P.C.$ It is not a simple and convenient process to calculate the total deflections from the $P.C.$ to a series of points on the second curve. It may readily be shown that adding the chord deflection for the second curve to the total deflection for the $P.C.C.$ or $P.R.C.$ will yield an incorrect result. Resetting at the $P.C.C.$ or $P.R.C.$ is quite simple, and the process of running in the second curve is similar in principle to that of § 59, page 31.

Compound Curves. 53

102. Data Used in Compound Curve Formulas.

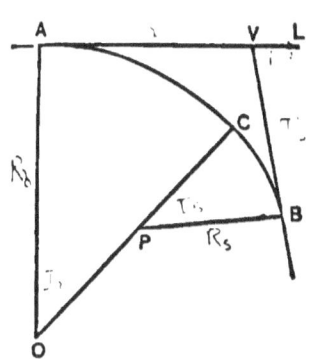

In the curve of larger radius,

$$OA = R_l$$
$$AOC = I_l$$
$$AV = T_l$$

In the curve of shorter radius,

$$PB = R_s$$
$$BPC = I_s$$
$$VB = T_s$$

also $\quad LVB = I$

103. Problem. *Given* R_l, R_s, I_l, I_s.
Required I, T_l, T_s.

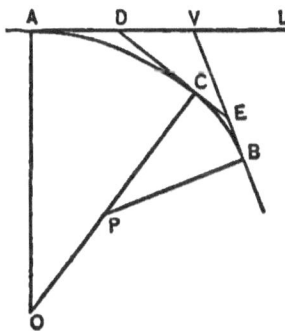

Draw the common tangent DCE.

Then $\quad I = I_l + I_s$
$$AD = CD = R_l \tan \tfrac{1}{2} I_l$$
$$EB = CE = R_s \tan \tfrac{1}{2} I_s$$

or find CD and CE using Searles' Table VI. and the Correction, Table V.

In the triangle DVE we have one side and three angles

$$DE = R_l \tan \tfrac{1}{2} I_l + R_s \tan \tfrac{1}{2} I_s \quad (45)$$
$$VDE = I_l$$
$$VED = I_s$$
$$DVE = 180 - I$$

Solve for VD and VE

$$AV = AD + VD = T_l$$
$$VB = BE + VE = T_s$$

This problem will be of use in making calculations for platting, in cases where the curves have been run in without finding V on the ground. The points D and E will seldom be fixed on the ground, and in cases where V is set, the problem is likely to take some form shown in one of the problems following.

104. Problem. *Given T_s, R_s, R_l, I.*
Required T_l, I_l, I_s.

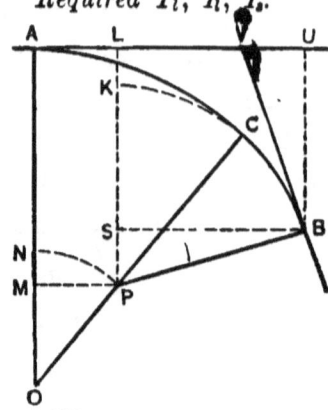

Draw arcs NP and KC.

Draw perpendiculars MP, LP, SB, UB.

Then
$$AM = LP$$
$$AN = R_s = KP$$
$$NM = LK = LS - KS$$
$$OP \text{ vers } NOP = VB \sin VBS - PB \text{ vers } KPB$$
$$(R_l - R_s) \text{ vers } I_l = T_s \sin I - R_s \text{ vers } I$$

$$\text{vers } I_l = \frac{T_s \sin I - R_s \text{ vers } I}{R_l - R_s} \tag{46}$$

$$I_s = I - I_l$$

$$AV = MP + SB - UV$$
$$T_l = (R_l - R_s) \sin I_l + R_s \sin I - T_s \cos I \tag{47}$$

105. Problem. *Given T_s, R_s, I_s, I.*
Required T_l, R_l, I_l.

$$I_l = I - I_s$$

$$R_l - R_s = \frac{T_s \sin I - R_s \text{ vers } I}{\text{vers } I_l} \tag{48}$$

$$T_l = (R_l - R_s) \sin I_l + R_s \sin I - T_s \cos I \tag{49}$$

106. Problem. *Given T_l, T_s, R_s, I.*
Required R_l, I_l, I_s.

$$\tan \tfrac{1}{2} I_l = \frac{T_s \sin I - R_s \text{ vers } I}{T_l + T_s \cos I - R_s \sin I} \tag{50}$$

$$R_l - R_s = \frac{T_l + T_s \cos I - R_s \sin I}{\sin I_l} \tag{51}$$

Compound Curves.

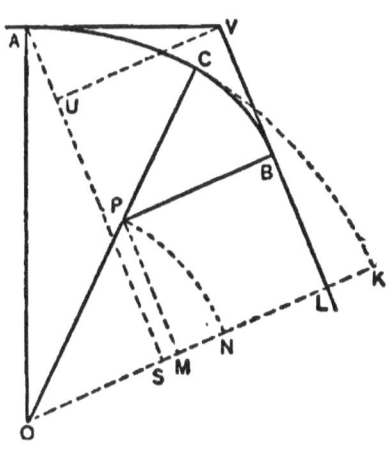

107. Problem.

Given T_l, R_l, R_s, I.
Required T_s, I_l, I_s.

Draw arcs NP, KC.

Draw perpendiculars OK, AS, PM, VU.

Then $LM = BP$
$\quad\quad\quad = KN$

$MN = LM - LN$
$\quad\quad = KN - LN$
$\quad\quad = KL$

$LK = MN = \quad KS \quad - \quad LS$
$OP \text{ vers } NOP = AO \text{ vers } AOK - AV \sin VAS$
$(R_l - R_s) \text{ vers } I_s = R_l \text{ vers } I - T_l \sin I$

$$\text{vers } I_s = \frac{R_l \text{ vers } I - T_l \sin I}{R_l - R_s} \quad (52)$$

$$I_l = I - I_s$$

$VB = \quad AS \quad - \quad PM \quad - \quad AU$
$T_s = R_l \sin I - (R_l - R_s) \sin I_s - T_l \cos I \quad (53)$

108. Problem. Given T_l, R_l, I_l, I.
Required T_s, R_s, I_s.

$$I_s = I - I_l$$

$$R_l - R_s = \frac{R_l \text{ vers } I - T_l \sin I}{\text{vers } I_s} \quad (54)$$

$$T_s = R_l \sin I - (R_l - R_s) \sin I_s - T_l \cos I \quad (55)$$

109. Problem. Given T_l, T_s, R_l, I.
Required R_s, I_l, I_s.

$$\tan \tfrac{1}{2} I_s = \frac{R_l \text{ vers } I - T_l \sin I}{R_l \sin I - T_l \cos I - T_s} \quad (56)$$

$$R_l - R_s = \frac{R_l \sin I - T_l \cos I - T_s}{\sin I_s} \quad (57)$$

110. Problem. *Given, in the figure,* AB, VAB, VBA, R_s.
Required R_l, I_l, I_s, I.

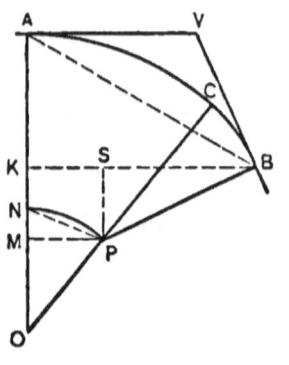

Draw arc NP; also perpendicular KB, MP, SP.

$I = \text{VAB} + \text{VBA}$

$$\begin{aligned}
\text{NM} &= \text{AK} + \text{KM} - \text{AN} \\
&= \text{AB}\sin\text{VAB} + \text{PB}\cos\text{SPB} - \text{AN} \\
&= \text{AB}\sin\text{VAB} + R_s \cos I - R_s \\
&= \text{AB}\sin\text{VAB} - R_s \operatorname{vers} I
\end{aligned}$$

$$\begin{aligned}
\text{MP} &= \text{KB} - \text{SB} \\
&= \text{AB}\cos\text{VAB} - \text{PB}\sin\text{SPB} \\
&= \text{AB}\cos\text{VAB} - R_s \sin I
\end{aligned}$$

$$\tan\text{NPM} = \tan \tfrac{1}{2} I_l = \frac{\text{NM}}{\text{MP}} \tag{58}$$

$$I_s = I - I_l$$

$$\text{OP} = R_l - R_s = \frac{\text{MP}}{\sin I_l} \tag{59}$$

111. Problem. *Given, in the figure,* AB, VAB, VBA, R_l.
Required R_s, I_l, I_s, I.

Find I, and show that

$$\tan \tfrac{1}{2} I_s = \frac{R_l \operatorname{vers} I - \text{AB}\sin\text{VBA}}{R_l \sin I - \text{AB}\cos\text{VBA}} \tag{60}$$

$$R_l - R_s = \frac{R_l \sin I - \text{AB}\cos\text{VBA}}{\sin I_s} \tag{61}$$

112. Problem. *Given a Simple Curve ending in a given tangent.*

A second curve of given radius is to leave this and end in a parallel tangent.

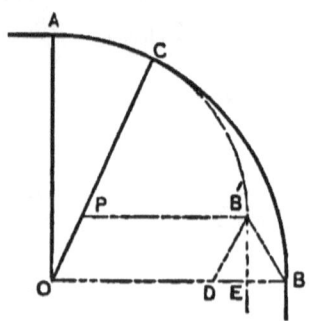

Required the P.C.C.

Let AB be the given curve of radius R_l.

C be the P.C.C.

CB' the second curve of radius R_s.

BE = p = distance between tangents.

Then $\operatorname{vers}\text{COB} = \dfrac{p}{R_l - R_s}$ (62)

Compound Curves. 57

It may sometimes be more convenient or quicker to run in a simple curve first and change to a compound curve by the method of this problem, rather than to run in the compound curve at first. When it is impossible or inconvenient to run in the curve as far as B ($P.T.$), the point of intersection D between the curve and the tangent may be found, the angle LDN measured, and BE calculated. In the figure below

$$BE = DO \text{ vers } DOB$$

$$p = R_t \text{ vers } LDN \qquad (63)$$

113. Example. *Given Notes of Curve.*

$$22 + 20 \; P.C. \qquad 5° \text{ curve } R$$

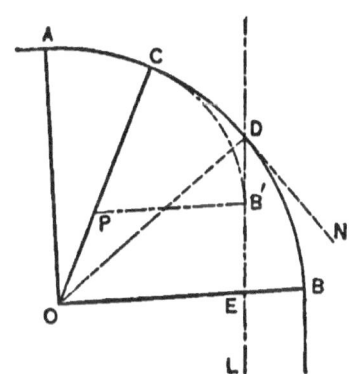

Proposed tangent intersects curve at $26 + 90$.

Angle between tangent and curve $= 10° 20'$.

Required Station of P.C.C. to join proposed tangent.

$$p = R_5 \text{ vers } 10° 20'$$

R_5 log 3.059290
10° 20' vers 8.210028
―――――――――――
p log 1.269318

vers COB $= \dfrac{p}{R_5 - R_7}$

$R_5 = 1146.28$
$R_7 = 819.02$
―――――
327.26

p log 1.269318
log 2.514893
―――――――
vers 8.754425

$26 + 90$
$1 + 81.3$
―――――
$25 + 08.7$ $P.C.C.$

$19° 24'$
$10° 20'$
―――――
$9° 04'$

$=$

$9.0607(5°$
―――――
181.3

114. Problem. *Given a Compound Curve ending in a tangent.*

Required to change the P.C.C. so as to end in a given parallel tangent, the radii remaining unchanged.

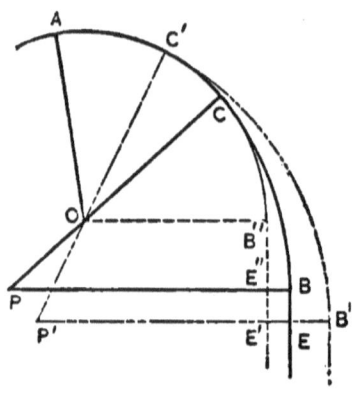

I. When the *new tangent* lies *outside* the *old tangent*, and the curve ends with curve of *larger radius*.

Let ACB be the given compound curve.

AC'B' the required curve.

Produce arc AC to B''.

Draw OB'' parallel to PB, and B''E' perpendicular to PB.

Let $B'E = p =$ perpendicular distance between tangents.

Then $B'E = B'E' - BE''$

$B'E = (R_l - R_s) \text{ vers } C'P'B' - (R_l - R_s) \text{ vers } COB''$

$p = (R_l - R_s) \text{ vers } I_l' - (R_l - R_s) \text{ vers } I_l$

$$\text{vers } I_l' = \text{vers } I_l + \frac{p}{R_l - R_s} \qquad (64)$$

$AOC' = I - I_l'$

II. When the *new tangent* lies *inside* the *old tangent*, and the *curve ends* with the curve of *larger radius*.

$$\text{vers } I_l' = \text{vers } I_l - \frac{p}{R_l - R_s} \qquad (65)$$

III. When the *new tangent* lies *outside* the *old tangent*, and the *curve ends* with curve of *smaller radius*.

$$\text{vers } I_s' = \text{vers } I_s - \frac{p}{R_l - R_s} \qquad (66)$$

IV. When the *new tangent* lies *inside* the *old tangent*, and the *curve ends* with curve of *smaller radius*.

$$\text{vers } I_s' = \text{vers } I_s + \frac{p}{R_l - R_s} \qquad (67)$$

Compound Curves. 59

115. Problem. *Given a Simple Curve joining two tangents. Required to substitute a symmetrical curve with flattened ends, using the same P.C. and P.T.*

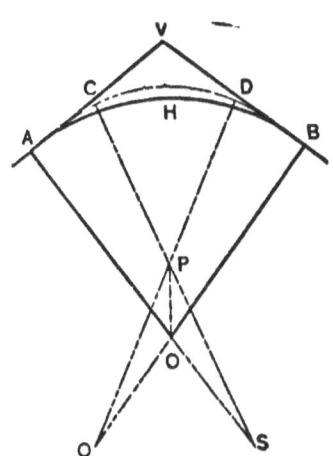

Let AHB be the simple curve of radius R_c.

ACDB the required curve in which

$$BQ = AS = R_l$$
$$PC = R_s$$
$$ASC = BQD = I_t$$
$$CPD = I_s$$

Then $\qquad I = I_s + 2 I_t$

In the triangle POQ,

$PQ = R_l - R_s$ $\qquad$ $PQO = I_t$
$QO = R_l - R_c$ $\qquad$ $POQ = 180 - \tfrac{1}{2} I$
$\qquad\qquad\qquad$ $OPQ = \tfrac{1}{2} I_s$

There are then *Given* I, R_c. *Required* R_l, R_s, I_t, I_s.

We may assume any two of the latter (except I_t and I_s), and readily calculate the others.

I. *Assume R_l and I_t.*

$$I_s = I - 2 I_t$$
$$PQ \;:\; QO \;= \sin POQ : \sin OPQ$$
$$R_l - R_s : R_l - R_c = \sin \tfrac{1}{2} I \;:\; \sin \tfrac{1}{2} I_s$$
$$R_l - R_s = \frac{(R_l - R_c) \sin \tfrac{1}{2} I}{\sin \tfrac{1}{2} I_s} \qquad (68)$$

II. *Assume R_l and R_s.*

$$\sin \tfrac{1}{2} I_s = \frac{(R_l - R_c) \sin \tfrac{1}{2} I}{R_l - R_s} \qquad (69)$$

60 Railroad Curves and Earthwork.

III. *Assume R_s and I_s.*

$$I_t = \tfrac{1}{2}(I - I_s)$$

$$\frac{PQ + QO}{PQ - QO} = \frac{\tan \tfrac{1}{2}(POQ + OPQ)}{\tan \tfrac{1}{2}(POQ - OPQ)}$$

$$\frac{R_l - R_s + R_l - R_c}{R_l - R_s - R_l + R_c} = \frac{\tan \tfrac{1}{2}(180 - \tfrac{1}{2}I + \tfrac{1}{2}I_s)}{\tan \tfrac{1}{2}(180 - \tfrac{1}{2}I - \tfrac{1}{2}I_s)}$$

$$2R_l - R_c - R_s = \frac{(R_c - R_s)\cot \tfrac{1}{4}(I - I_s)}{\cot \tfrac{1}{4}(I + I_s)} \qquad (70)$$

116. Problem. *Given a Simple Curve joining two tangents. Required to substitute a curve with flattened ends to pass through the same middle point.*

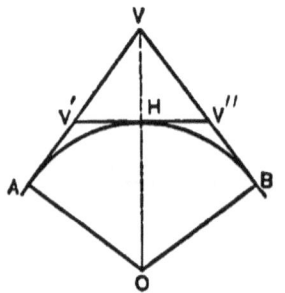

Let AB be the given simple curve, and H the middle point.

Erect an auxiliary tangent V'HV'' at H.

The auxiliary intersection angles at V' and V'' are readily calculated; also V'H and V'''H.

Sufficient additional data can be assumed, and the problem solved as a problem in compound curves.

It is not *necessary* that the curves on the two sides of H should be symmetrical.

Having given   V'H = T_s

   VV'H = I

Assume $R_l >$ AO

   $R_s <$ AO

and apply (46) and (47).

Other assumptions of similar sort will allow the use of other formulas on pages 54 and 55.

Both this and the preceding problem will be found of considerable value in revising the alignment of track, and introducing flatter ends for the curves so that the transition from tangent to curve shall be less abrupt.

Compound Curves.

117. Problem. *Given two simple curves with connecting tangent.*

Required to substitute a *simple curve* of *given radius* to connect the two.

Let $DC = l =$ the given tangent, connecting the two curves AD and CB, of radii R_s and R_l, respectively.

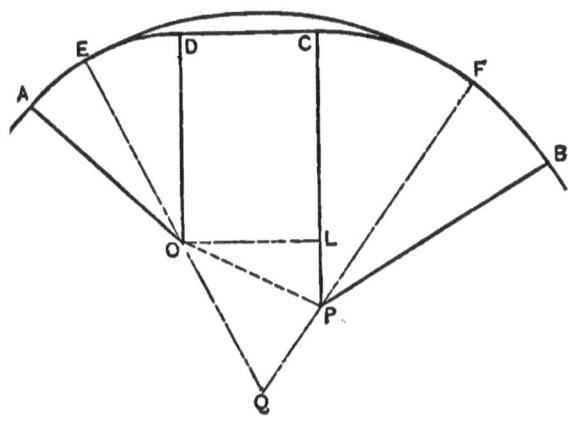

Let EF be required curve of radius R_c.
Join OP, and draw perpendicular OL.

Then
$$\tan LOP = \frac{LP}{OL} = \frac{R_l - R_s}{l}$$

$$OP = \frac{l}{\cos LOP}$$

In the triangle OPQ we have given

$$OP = \frac{l}{\cos LOP}; \quad OQ = R_c - R_s; \quad QP = R_c - R_l$$

Solve this triangle for OQP, QOP, OPQ

Then
$$CPF = 180° - (OPQ + OPL)$$
$$EOD = 90° - (QOP + LOP)$$

CHAPTER VI.

REVERSED CURVES.

118. It is very undesirable that reversed curves should be used on main lines, or where trains are to be run at any considerable speed. The marked change in direction is objectionable, and an especial difficulty results from there being no opportunity to elevate the outer rail at the $P.R.C.$ The use of reversed curves on lines of railroad is therefore very generally condemned by engineers. For yards and stations, reversed curves may often be used to advantage, also for street railways, and perhaps for other purposes.

119. Problem. *Given the perpendicular distance between parallel tangents, and the common radius of the reversed curve.*
Required the central angle of each curve.

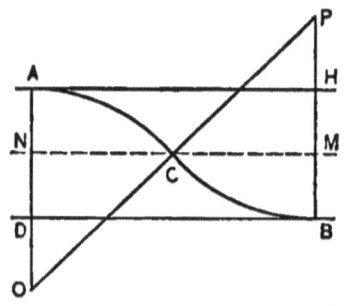

Let AH and BD be the parallel tangents.

ACB the reversed curve.

HB = p = perpendicular distance between tangents.

Draw perpendicular NM.

Let AOC = BPC = I_r.

Then $\quad$ vers AOC = $\dfrac{AN}{AO} = \dfrac{BM}{PB} = \dfrac{\frac{1}{2}HB}{AO}$

$$\text{vers } I_r = \frac{\frac{1}{2}p}{R} \qquad (71)$$

120. Problem. *Given p, I_r.*
Required R.

$$R = \frac{\frac{1}{2}p}{\text{vers } I_r} \qquad (72)$$

Reversed Curves. 63

121. Problem. *Given the perpendicular distance between parallel tangents, and chord distance between P.C. and P.T.*

Required the common radius of reversed curve to connect the parallel tangents.

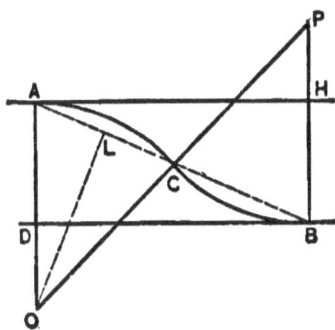

Let AH and BD be the parallel tangents.
ACB the reversed curve.
BH $= p$
AB $= d$
Connect AC and CB.
Then AOC $=$ BPC, and
ACO $=$ PCB
ACB is a straight line
AO : AL $=$ AB : HB
$$R : \frac{d}{4} = d : p$$
$$R = \frac{d^2}{4p} \qquad (73)$$

122. Problem. *Given R and p.*
Required d.
$$d = \sqrt{4\,Rp} = 2\sqrt{Rp}. \qquad (74)$$

123. Problem. *Given the perpendicular distance between two parallel tangents, and the central angle and radius of first curve of reversed curve.*

Required the radius of second curve.

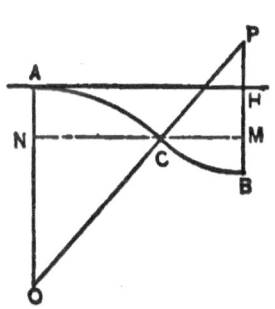

Let ACB $=$ reversed curve
HB $= p$
AO $= R_1$
AOC $=$ CPB $= I_r$
PB $= R_2$
Draw perpendicular NCM.
HB $=$ AN $+$ MB
   $=$ AO vers AOC $+$ BP vers BPC
$p = R_1$ vers I_r $+ R_2$ vers I_r

$$R_1 + R_2 = \frac{p}{\text{vers } I_r} \qquad (75)$$

124. Problem. *Given R_1, R_2, p.*
Required I_r.

$$\operatorname{vers} I_r = \frac{p}{R_1 + R_2} \tag{76}$$

125. Problem. *Given two points upon tangents not parallel, the length of line joining the two points, and the angles made by this line with each tangent.*
Required the common radius of a reversed curve to connect the two tangents at the given points.

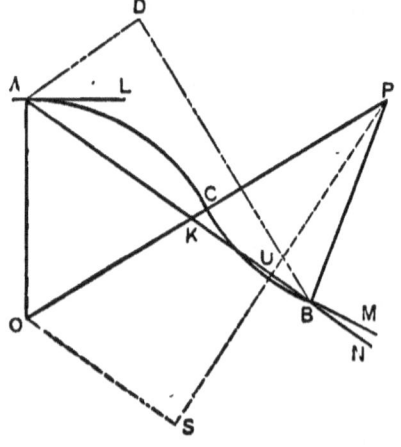

Let A and B be the given points.

AL, BM = given tangents
ACB = required curve
LAB = A and MBN = B
AB = l
AKO = PKB = K

Draw PS perpendicular and OS parallel to AB.
Also BD perpendicular and AD parallel to OP.

Then PS = PU + SU

OP sin POS = PB cos BPU + AO sin OAB

$2R \sin K = R \cos B + R \cos A$

$$\sin K = \frac{\cos A + \cos B}{2}$$

AOK = O = 180° − OKA − OAK
  = 180° − K − (90° − A) = 90° + A − K

BPK = P = 180° − BKP − PBK =
  = 180° − K − (90° − B) = 90° + B − K

BD = AB sin DAB = AO sin AOK + BP sin KPB

$l \sin K = R \sin O + R \sin P$

$$R = \frac{l \sin K}{\sin O + \sin P} \tag{77}$$

Reversed Curves.

126. Problem. *Given the length of the common tangent and the angles of intersection with the separated tangents.*

Required the common radius of a reversed curve to join the two separated tangents.

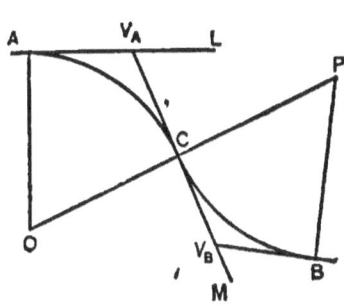

Let $V_A V_B$ = common tangent
AV_A, BV_B = separated tangents
ACB = required curve
$LV_A C = I_A$
$MV_B B = I_B$
$V_A V_B = l$
$V_A V_B = V_A C + V_B C$
$l = R \tan \tfrac{1}{2} I_A + R \tan \tfrac{1}{2} I_B$

$$R = \frac{l}{\tan \tfrac{1}{2} I_A + \tan \tfrac{1}{2} I_B} \quad (78)$$

An approximate method is as follows: —
Find T'_{A1} for a 1° curve; also T_{B1} (Table VI., Searles).
Then
$$D_a = \frac{T_{A1} + T_{B1}}{V_A V_B}$$

127. Problem. *Given a P.C. upon one of two tangents not parallel, also the tangent distance from P.C. to V, also the angle of intersection, also the unequal radii of a reversed curve to connect the tangents.*

Required the central angles of the simple curves, and tangent distance, V to P.T.

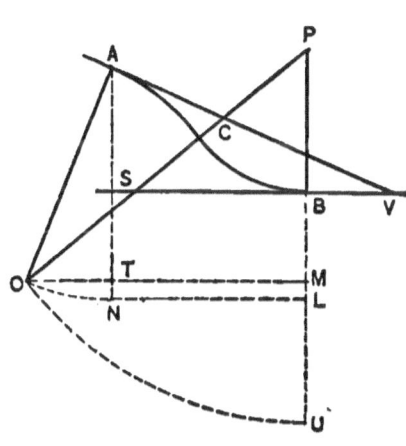

Let $AV = T_1$ = given tangent distance.
A = given P.C.
V = vertex
$AVS = I$
$AO = R_1$ } given radii
$PB = R_2$
VS = second tangent
ACB = required curve
$AOC = I_1$ } required angles
$BPC = I_2$
$BV = T_2$ = required tangent distance.

66 Railroad Curves and Earthwork.

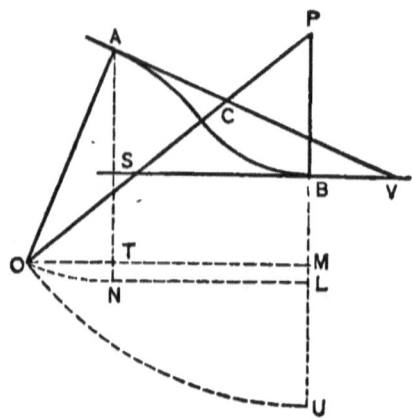

Draw arcs OU, ON, also perpendiculars AN, OM, NL.

Then AO = BU = R_1

LU = BU − BL = AN − SN = AS

LU = AS = MU − ML

AV sin AVS = PO vers BPC − AO vers OAN

$T_1 \sin I = (R_1 + R_2) \text{ vers } I_2 - R_1 \text{ vers } I$

$$\text{vers } I_2 = \frac{R_1 \text{ vers } I + T_1 \sin I}{R_1 + R_2} \qquad (79)$$

$$I_1 = I_2 - I$$

$$T_2 = T_1 \cos I + R_1 \sin I - (R_1 + R_2) \sin I_2 \qquad (80)$$

128. Problem. *Given* BV *instead of* AV, *and other data as above.*

Required I_1, I_2, *etc.*

Draw a new figure similar in principle to the preceding and solve in a similar way, and using the same notation as above

$$\text{vers } I_1 = \frac{R_2 \text{ vers } I + T_2 \sin I}{R_1 + R_2} \qquad (81)$$

$$T_1 = T_2 \cos I + R_2 \sin I + (R_1 + R_2) \sin I_1 \qquad (82)$$

CHAPTER VII.

PARABOLIC CURVES.

129. Instead of circular arcs to join two tangents, parabolic arcs have been proposed and used, in order to do away with the sudden changes in direction which occur where a circular curve leaves or joins a tangent. Parabolic curves have, however, failed to meet with favor for railroad curves for several reasons.

1. Parabolic curves are less readily laid out by instrument than are circular curves.

2. It is not easy to compute at any given point the radius of curvature for a parabolic curve; it may be necessary to do this either for curving rails or for determining the proper elevation for the outer rail.

3. The use of the "Spiral," or other "Easement," or "Transition" curves secures the desired result in a more satisfactory way.

There are however many cases (in Landscape Gardening or elsewhere) where a parabolic curve may be useful either because it is more graceful or because, without instrument, it is more easily laid out, or for some other reason.

It is seldom that parabolic curves would be laid out by instrument.

130. Properties of the Parabola.

(*a*)* The locus of the middle points of a system of parallel chords of a parabola is a straight line parallel to the axis of the parabola (*i.e.* a diameter).

(*b*) The locus of the intersection of pairs of tangents is in the diameter.

(*c*) The tangent to the parabola at the vertex of the diameter is parallel to the chord bisected by this diameter.

(*d*) Diameters are parallel to the axis.

---

* §§ 131, 132, 133, 134, RUNKLE'S *Analytical Geometry*.

(e) The equation of the parabola, the coördinates measured upon the diameter and the tangent at the end of the diameter is

$$y'^2 = \frac{4p}{\sin^2 \theta} x'$$

or
$$y^2 = 4p'x \qquad (83)$$

131. Problem. *Given two tangents to a parabola, also the position of P.C. and P.T.*
Required to lay out the parabola by "offsets from the tangent."

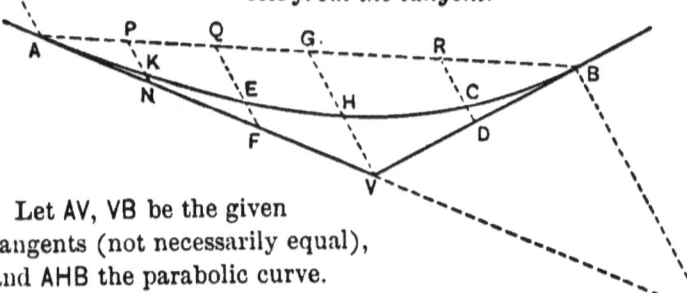

Let AV, VB be the given tangents (not necessarily equal), and AHB the parabolic curve.
Join the chord AB.
Draw VG bisecting AB.
Draw AX, BY, parallel to VG.
Produce AV to Y.
Then VG is a diameter of the parabola.
    AX parallel to VG is also a diameter.
The equation of the parabola referred to AX and AY as axes is

$$y^2 = 4p'x.$$

Hence   AV² : AY² = HV : BY

AV² : (2 AV)² = HV : 2 GV

1 : 4 = HV : 2 GV

$$HV = \frac{GV}{2} \qquad (84)$$

Next bisect VB at D.
Draw CD parallel to AX.

Then   BD² : BV² = CD : HV

$$CD = \frac{HV}{4}$$

Parabolic Curves.

Similarly, make $\quad AN = NF = FV$

Then $\quad KN = \dfrac{HV}{9}$

$$EF = \dfrac{4}{9}HV$$

In a similar way, as many points as are needed may be found.

132. Field-work.

(a) Find G bisecting AB.

(b) Find H bisecting GV.

(c) Find points P, Q, and N, F, dividing AG, AV, proportionately; also R and D, dividing GB and BV proportionately.
Use simple ratios when possible (as $\frac{1}{2}$, $\frac{1}{3}$, etc.).

(d) Lay off on PN, the calculated distance KN

on QF lay off EF

on RD lay off CD

In figure opposite, $\quad KN = \dfrac{HV}{9}$

$$CD = \dfrac{HV}{4}$$

$$EF = \dfrac{4}{9}HV$$

For many purposes, or in many cases, it will give results sufficiently close, to proceed without establishing P, Q, R; the directions of NK, EF, CD, being given approximately by eye. When the angle AVG is small (as in the figure), it will generally be necessary to find P, Q, R, and fix the directions in which to measure NK, EF, CD. When the angle AVG is large (greater than 60°) and the distances NK, EF, CD are not large, it will often be unnecessary to do this. No fixed rule can be given as to when approximate methods shall be used. Experience educates the judgment so that each case is settled upon its merits.

133. Problem. *Given two tangents to a parabola, also the positions of P.C. and P.T.*

Required to lay out the parabola by " middle ordinates."

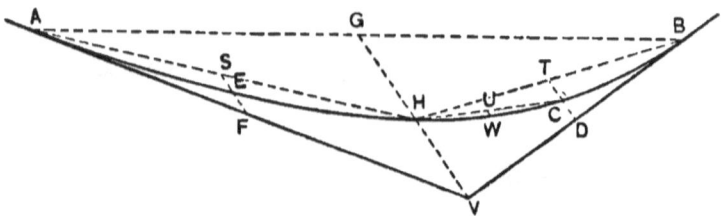

The ordinates are taken from the middle of the chord, and parallel to GV in all cases.

Field-work.

(*a*) Establish H as in last problem.
(*b*) Lay off SE = ¼ HV; also TC = ¼ HV.
(*c*) Lay off UW = ¼ TC, and continue thus until a sufficient number of points is obtained.

The length of curve can be conveniently found only by measurement on the ground.

Note the difference in method between § 85 and § 133.

134. Vertical Curves.

It is convenient and customary to fix the grade line upon the profile as a succession of straight lines ; also to mark the elevation above datum plane of each point where a change of grade occurs ; also to mark the rates of grade in feet per station of 100 feet. At each change of grade a vertical angle is formed. To avoid a sudden change of direction it is customary to introduce a vertical curve at every such point where the angle is large enough to warrant it. The curve commonly used for this purpose is the parabola. A circle and a parabola would substantially coincide where used for vertical curves. The parabola effects the transition rather better theoretically than the circle, but its selection for the purpose is due principally to its greater simplicity of application. It is generally laid to extend an equal number of stations on each side of the vertex.

Parabolic Curves. 71

135. Problem. *Given the elevations at the vertex and at one station (100') each side of vertex.*
Required the elevation of the vertical curve opposite the vertex.

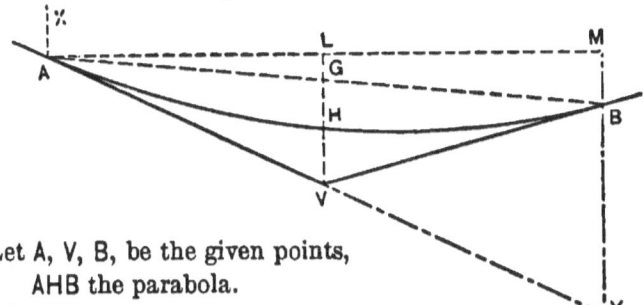

Let A, V, B, be the given points,
  AHB the parabola.
Join AB.
Draw vertical lines AX, LGHV, MBY, and horizontal line ALM.
Produce AV to Y.

In the case of a vertical curve, the horizontal projections of AV and VB are equal, and here each equals 100 feet = AL = ML.

Therefore   AG = GB, and AV = VY

VG is a diameter of the parabola.
AX is also a diameter.

$$VH = \frac{VG}{2}$$

$$\text{Elev. } H = \frac{1}{2}\left(\frac{\text{Elev. A} + \text{Elev. B}}{2} + \text{Elev. V}\right) \quad (85)$$

This affords a simple and quick method of finding H when the vertical curve extends only one station each side of vertex, which is the most common case. Other methods or rules for vertical curves are used on various railroads, or by different engineers, and it will prove interesting and valuable to investigate such rules, to discover whether the resulting curve is a parabola, as will generally be found to be the case. When the vertical curve extends more than one station each side of the vertex, the following method is preferable, which is also applicable to the above case, and is in some respects preferable for that also.

72 Railroad Curves and Earthwork.

136. Problem. *Given the rates of grade g of AV; g' of VB; the number of stations n, on each side of vertex, covered by the vertical curve; also the elevation of the point A.*
Required the elevation, at each station, of the parabola AB.

Draw vertical lines

    DD'D", EE'E", VHL, YBM

Also horizontal lines

    VC, ALM

Produce AV to Y

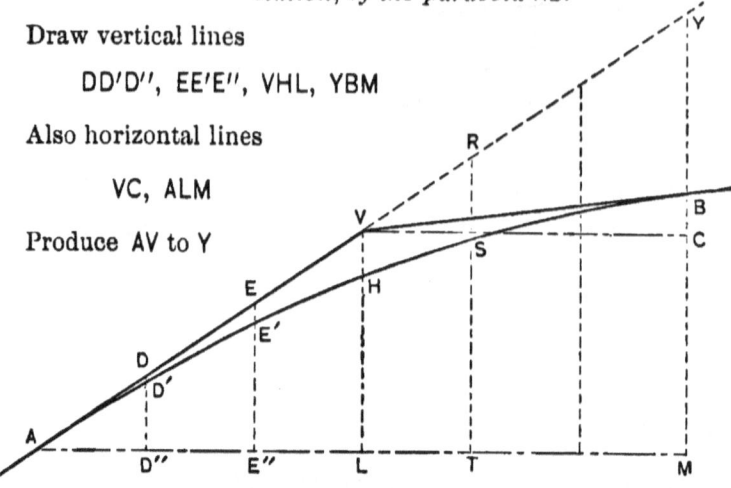

Let a_1 = offset DD' at the first station from A.

       a_2 =   "   EE'   "   second   "   " A.

       a_3 = etc.

Then $a_2 = 2^2 a_1 = 4 a_1$

          $a_3 = 3^2 a_1 = 9 a_1$

          $a_{2n} = (2n)^2 a_1 = 4 n^2 a_1 = $ YB

          YB =   YC − BC

      $4 n^2 a_1 = $ $ng - ng'$

$$a_1 = \frac{g - g'}{4n} \qquad (86)$$

Due regard must be given to the signs of g and g' in this formula, whether + or −.

Parabolic Curves.

From the elevation at A we may now find the required elevations, since we have given g,

and we also have $\quad a_1$

$$a^2 = 4\, a_1$$

$$a_3 = 9\, a_1 \text{ etc.}$$

A method better and more convenient for use is given below.

$$\text{DD}'' = g\,; \quad \text{D}'\text{D}'' = g - a_1$$

$$\text{EE}'' = 2\,g\,; \quad \text{E}'\text{E}'' = 2\,g - a_2 = 2\,g - 4\,a_1$$

$$\text{VL} = 3\,g\,; \quad \text{HL} = 3\,g - a_3 = 3\,g - 9\,a_1$$

$$\text{RT} = 4\,g\,; \quad \text{TS} = 4\,g - a_4 = 4\,g - 16\,a_1 \text{ etc.}$$

Again, $\quad \text{D}'\text{D}'' = g - a_1 \qquad\qquad = g - a_1$

$$\text{E}'\text{E}'' - \text{D}'\text{D}'' = 2\,g - 4\,a_1 - (\,g - a_1) = g - 3\,a_1$$

$$\text{HL} - \text{E}'\text{E}'' = 3\,g - 9\,a_1 - (2\,g - 4\,a_1) = g - 5\,a_1$$

$$\text{ST} - \text{HL} = 4\,g - 16\,a_1 - (3\,g - 9\,a_1) = g - 7\,a_1 \text{ etc.}$$

On a straight grade, the elevation of any station is found from the preceding, by adding a constant g.

On a vertical curve, the elevation of each station is found from the preceding by adding, in a similar way, not a constant, but a varying increment, being for the

1st station from $A = g - a_1$ ⎤
2d   "   " $A = g - 3\,a_1$ ⎬ changing by successive differences of $2\,a_1$ in each case.
3d   "   " $A = g - 5\,a_1$ ⎦

The labor involved is not materially greater in many cases, for a vertical curve than for a straight grade. This method has the additional advantage that a correct final result at the end of the vertical curve makes a "check" upon all intermediate results.

74   Railroad Curves and Earthwork.

137. Example.

Given. Grades as follows :—

Sta.	Elev.	Rate
5	117.50	
10	135.00	+ 3.50
15	137.50	+ 0.50

Then $a_1 = \dfrac{g - g'}{4\,n} = \dfrac{3.00}{12} = 0.25$

Sta.	Elev.		
5	117.50		
	+ 3.50		$= g$
6	121.00		
	+ 3.50	3.50	$= g$
7	124.50	− 0.25	$= a_1$
	+ 3.25	3.25	$= g - a_1$
8	127.75	− 0.50	$-2\,a_1$
	+ 2.75	2.75	$g - 3\,a_1$
9	130.50	− 0.50	$-2\,a_1$
	+ 2.25	2.25	$g - 5\,a_1$ etc.
10	132.75	− 0.50	
	+ 1.75	1.75	
11	134.50	− 0.50	
	+ 1.25	1.25	
12	135.75	− 0.50	
	+ 0.75	0.75	
13	136.50	End of curve	
	+ 0.50		$= g'$
14	137.00		
	+ 0.50		
15	137.50		

The elevation for Sta. 15 thus obtained agrees with the elevation shown in the data. All the intermediate elevations are therefore "checked."

138. Problem.   *Given* g, g', a_1
   Required n

From (86) $n = \dfrac{g - g'}{4\,a_1}$ (87)

From practical considerations a_1 should not exceed 0.25 or

$$n \not> \dfrac{g - g'}{4 \times 0.25} \quad \text{or} \quad n \not> g - g' \qquad (88)$$

CHAPTER VIII.

TURNOUTS.

139. A Turnout is a track leading from a main or other track.

Turnouts may be for several purposes.

I. *Branch Track* (for line used as a Branch Road for general traffic).

II. *Siding* (for passing trains at stations, storing cars, loading or unloading, and various purposes).

III. *Spur Track* (for purposes other than general traffic, as to a quarry or warehouse).

IV. *Cross Over* (for passing from one track to another, generally parallel).

The essential parts of a turnout are

1. *The Switch.*   2. *The Frog.*   3. *The Guard Rail.*

1. Some device is necessary to cause a train to turn from the main track; this is called the "*Switch.*"

2. Again, it is necessary that one rail of the turnout track should cross one rail of the main track; and some device is necessary to allow the flange of the wheel to pass this crossing; this device is called a "*Frog.*"

3. Finally, if the flange of the wheel were allowed to bear against the point of the frog, there is danger that the wheel might accidentally be turned to the wrong side of the frog point. Therefore a *Guard Rail* is set opposite to the frog, and this prevents the flange from bearing against the frog point.

Frogs are of various forms and makes, but are mostly of this general shape, and the parts are named as follows : —

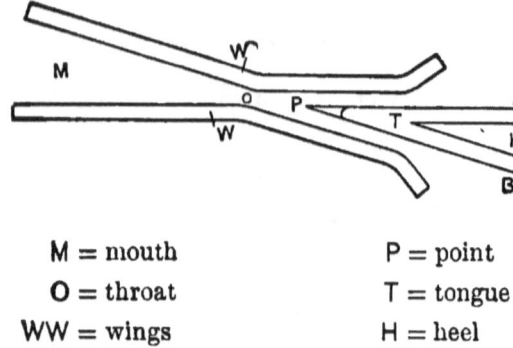

M = mouth   P = point
O = throat   T = tongue
WW = wings   H = heel

This shows the "stiff" frog.

The "spring" frog is often used where the traffic on the main line is large, and on the turnout small. In the spring frog W'W' is movable. AD represents the main line, and W'W'

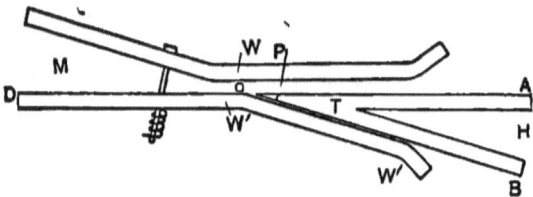

is pushed aside by the wheels of a train passing over the turnout. The "*Frog angle*" is the angle between the sides of the tongue of the frog = APB.

Frogs are made of certain standard proportions, and are classified by their number.

The "*Number*" n of a frog is found by dividing the length of the tongue by the width of the heel; that is, $n = \dfrac{PH}{AB}$.

140. Problem.   *Given n.   Required Frog Angle F.*

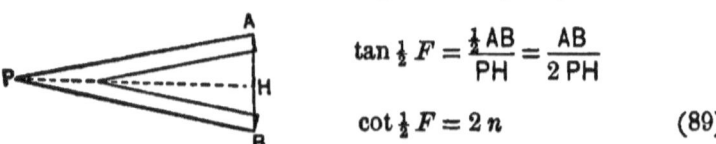

$$\tan \tfrac{1}{2} F = \frac{\tfrac{1}{2} AB}{PH} = \frac{AB}{2\,PH}$$

$$\cot \tfrac{1}{2} F = 2\,n \qquad\qquad (89)$$

Turnouts. 77

141. A form of switch in common use is the "*Stub-Switch*," which is formed by two rails, one on each side of the track, called the *Switch Rails*. One end of the rail for a short distance (often about 5 feet) is securely spiked to the ties, the rest of the rail being free to slide on the ties, so that it may meet the fixed rails of either main track or turnout, as desired. These fixed rails, supported on a *Head Block*, are held by a casting, or piece of metal called the *Head Chair*, and upon which the switch rail slides. A *Switch Rod* connects the ends of the switch rails with the *Switch Stand*. One end of the rail is spiked down, so that when the free end is drawn over by the switch rod, the rail is sprung into a curve which may with slight error be considered a circular curve, tangent to the main line (if this be straight). The distance through which the free end of the rail is drawn or *Thrown* by the switch rod is called the *Throw* of the switch. The free end of the rail is called the *Toe*, and the $P.C.$ of the curve the *Heel* of the switch.

Knowing the throw t and the length l of the switch rail, we can deduce the radius R, or degree of curve D, and continuing this curve to the point of frog, we can readily deduce the angle between the rails or the *Frog Angle* necessary.

It is more customary, however, having given the *throw* ($5''-5\frac{1}{2}''-6''$ are used on different roads) to assume either

(1) the radius (or degree) of turnout curve, and from this find F (or n) and l; or

(2) the number n (or angle F) of frog, and from this find R and then l.

142. When there are two turnouts at the same point, one on each side of the main line, three frogs are necessary, the middle one being called the "*Crotch Frog*."

It is necessary that there should be two numbers of frog, one for the ordinary turnout frog and another for the "crotch frog." It is advisable that only two be used, and that all turnout curves be arranged to use one or the other of these two frogs. (For double turnouts with point switches a third number may be necessary for a crotch frog. In this case we should assume n, or F, as the value for one of the two *standard frogs*.)

On main line it is now customary to use a "*Split Switch*" or "*Point Switch*," the description and discussion of which will follow the discussion of the stub switch.

In the figure the

Heel of Switch is at E or Q		Length of Switch Rail	=	QH
Toe of Switch	H	Throw of Switch	=	HI
Head Block	H and I	Lead	=	FI
Crotch Frog	C	Frog is at		F
Center of turnout curve	AB			

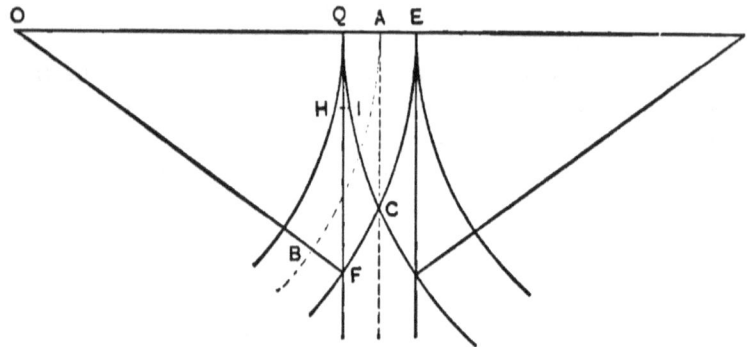

143. Problem. *Given gauge of track, g, frog angle, F, and throw of switch, t.*

Required R, l, and $QF = E$.

Let EF, QR, be the rails of the turnout.
Draw perpendicular KF. Also HI at toe of switch.

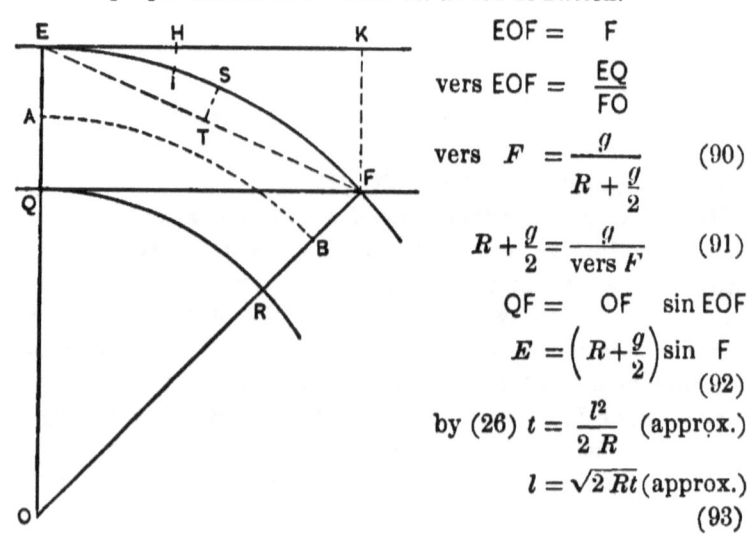

$$EOF = F$$

$$\text{vers } EOF = \frac{EQ}{FO}$$

$$\text{vers } F = \frac{g}{R + \frac{g}{2}} \quad (90)$$

$$R + \frac{g}{2} = \frac{g}{\text{vers } F} \quad (91)$$

$$QF = OF \sin EOF$$

$$E = \left(R + \frac{g}{2}\right) \sin F \quad (92)$$

by (26) $t = \dfrac{l^2}{2R}$ (approx.)

$l = \sqrt{2Rt}$ (approx.)

$$(93)$$

Turnouts.

144. Problem.   *Given g, t, n.*
   Required R, E, l.

In the figure preceding connect EF.

Then
$$EFQ = FEK = \tfrac{1}{2} F$$
$$QF = EQ \cot EFQ$$
$$E = g \cot \tfrac{1}{2} F \qquad (94)$$
$$= 2\,ng \qquad (95)$$
$$FQ^2 = FO^2 - QO^2$$
$$E^2 = \left(R + \tfrac{g}{2}\right)^2 - \left(R - \tfrac{g}{2}\right)^2$$
$$= \left(R + \tfrac{g}{2} + R - \tfrac{g}{2}\right)\left(R + \tfrac{g}{2} - R + \tfrac{g}{2}\right)$$
$$E^2 = 2R \times g$$
$$R = \frac{E^2}{2g} = \frac{4g^2 n^2}{2g}$$
$$R = 2\,n^2 g \qquad (96)$$
$$l^2 = E^2 \frac{t}{g}$$
$$= 4\,n^2 g^2 \frac{t}{g} = 4\,n^2 gt$$
$$l = 2\,n\sqrt{gt} \qquad (97)$$

145. Problem.   *Given g.*
   Required the middle ordinate ST.

from (34) $$M = \frac{c^2}{8R}$$

EQ is middle ordinate for chord 2 FQ
ST "   "   "   "   "   "   EF
EF = FQ (approx.)
ST : EQ = EF^2 : (2 FQ)^2
$$ST = \frac{EQ}{4}$$
$$ST = \frac{g}{4} \qquad (98)$$

This is true evidently whatever the degree of curve.

146. Problem. *Given g, R.*
Required, the angle of crotch frog, C.

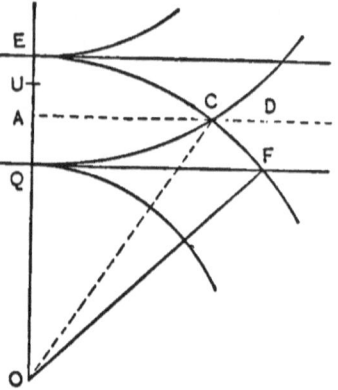

The frog angle at $C = 2\,AOC$

$$\text{vers } AOC = \frac{EA}{OC}$$

$$\text{vers } \tfrac{1}{2} C = \frac{\tfrac{1}{2} g}{R + \tfrac{g}{2}} \qquad (99)$$

147. Problem. *Given the number of crossing frog $= n_f$.*
Required the number of crotch frog $= n_c$.

$$AO = R = 2\,n_f^2 g$$

If we consider AD to represent a rail and n_q the frog proper for the crossing of QC and AD,

then $\qquad UO = 2\,n_q^2 \dfrac{g}{2}$

But the angle between EC and QC = twice the angle between QC and AD.

Then $\qquad UO = 2\,(2\,n_c)^2 \dfrac{g}{2}$ (approx.)

$$R = 2\,(2\,n_c)^2 \dfrac{g}{2} \quad \text{(approx.)}$$

$$2\,n_f^2 g = 2\,(2\,n_c)^2 \dfrac{g}{2} \quad \text{(approx.)}$$

$$n_f^2 = 2\,n_c^2 \quad \text{(approx.)}$$

$$0.7071\,n_f = n_c \quad \text{(approx.)} \qquad (100)$$

The approximation indicated here gives results much more nearly correct than it is practicable to work to in practice. For instance, in the case of a No. 9 frog, the number of the corresponding crotch frog is 6.359

By (100) it is   6.364
Railroads use   6.5

Turnouts.

148. Problem. *Given main track of radius R_m; also F, g, and n.*

Required radius R_t of a turnout inside of main track; also E.

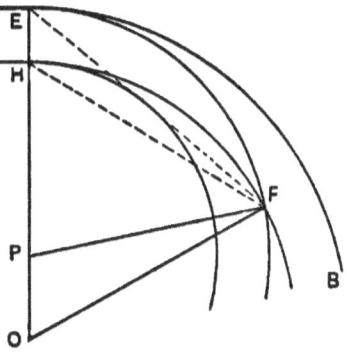

Let EB be the outer rail of main track.

EF the outer rail of turnout.

Join EF.

Let EOF $= O$; PFO $= F$.

In the triangle EOF, we have

$$EO = R_m + \frac{g}{2}$$

$$FO = R_m - \frac{g}{2}$$

$$EFO - FEO = EFO - EFP = F$$

$$EFO + FEO = 180° - O$$

Then $\tan \tfrac{1}{2}(\text{EFO}+\text{FEO}) : \tan \tfrac{1}{2}(\text{EFO}-\text{FEO}) = EO+FO : EO-FO$

$\cot \tfrac{1}{2} O$ $\quad\quad : \tan \tfrac{1}{2} F \quad = \quad 2 R_m \;:\; g$

$\tan \tfrac{1}{2} O$ $\quad\quad : \cot \tfrac{1}{2} F \quad = \quad g \;\;:\; 2 R_m$

$$\tan \tfrac{1}{2} O = \frac{g}{2 R_m} \cot \tfrac{1}{2} F = \frac{g\,2\,n}{2 R_m}$$

$$\tan \tfrac{1}{2} O = \frac{gn}{R_m} \tag{101}$$

Similarly, $\quad\quad\quad$ FPH $= F + O$

Join HF, and in triangle HPF

$$\tan \tfrac{1}{2}(F + O) = \frac{gn}{R_t}$$

$$R_t = \frac{gn}{\tan \tfrac{1}{2}(F + O)} \tag{102}$$

$$\text{chord HF} = E = 2\left(R_m - \frac{g}{2}\right)\sin \tfrac{1}{2} O \tag{103}$$

149. Approximate Formula.

Let R, D = Radius and a Degree of a *turnout curve from a straight line* to correspond to the given value of F or n.

R_m, D_m = Radius and Degree of *main track*.

R_t, D_t = Radius and Degree of *turnout curve*.

Then from (101) (102) $R_m = \dfrac{ng}{\tan \frac{1}{2} O}$; $R_t = \dfrac{ng}{\tan \frac{1}{2}(O + F)}$

also (96) $R = 2n^2 g = ng \times 2n = \dfrac{ng}{\tan \frac{1}{2} F}$

$$\sin \tfrac{1}{2} D_m = \frac{50}{R_m} = \frac{50 \tan \frac{1}{2} O}{ng}$$

$$\sin \tfrac{1}{2} D_t = \frac{50}{R_t} = \frac{50 \tan \frac{1}{2}(O + F)}{ng}$$

$$\sin \tfrac{1}{2} D = \frac{50}{R} = \frac{50 \tan \frac{1}{2} F}{ng}$$

$\sin \tfrac{1}{2} D_m : \sin \tfrac{1}{2} D_t : \sin \tfrac{1}{2} D = \tan \tfrac{1}{2} O : \tan \tfrac{1}{2}(O+F) : \tan \tfrac{1}{2} F$

$D_m : \quad D_t : \quad D = \quad O : \quad O+F \quad : F$ (approx.)

$D_m + D \quad : \quad D_t = \quad O+F : O+F \quad$ (approx.)

$$D_t = D_m + D \text{ (approx.)} \qquad (104)$$

Again, (103) $\mathsf{HF} = E = 2\left(R_m - \dfrac{g}{2}\right) \sin \tfrac{1}{2} O$

But $\dfrac{g}{2}$ is small compared with R_m, and may be neglected, and for small angles $\sin \tfrac{1}{2} O = \tan \tfrac{1}{2} O$ (approx.)

$$\mathsf{HF} = E = 2 R_m \tan \tfrac{1}{2} O \text{ (approx.)}$$

$$(101)\ E = 2gn \text{ (approx.)} \qquad (105)$$

This agrees with (95) $E = 2ng$ (turnout from straight track)

The above formula and (104), while approximate, are the formulas in general use.

It is difficult in practical track work to secure results more precise than would be obtained by the use of these approximate formulas.

Turnouts.

150. Problem. *Given main track of radius R_m, also F, g, n. Required radius R_t of a turnout curve outside of main track.*

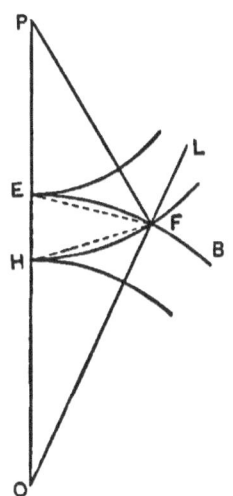

I. When the center of turnout curve lies *outside* of main track.

Let EB be the outer rail of main track, and HF the outer rail of turnout.

Join HF.

Let   HOF $= O$

Then   PFL $= F$

In the triangle HOF

$$FO = R_m + \frac{g}{2}$$

$$HO = R_m - \frac{g}{2}$$

Also   FHO $+$ HFO $= 180° - O$

FHO $-$ HFO $= 180° -$ PHF $- (180° -$ PHF $- F)$

                          $= F$

Then

$\tan \tfrac{1}{2}$(FHO $+$ HFO) : $\tan \tfrac{1}{2}$(FHO $-$ HFO) $=$ FO $+$ HO : FO $-$ HO

$\cot \tfrac{1}{2} O$: $\tan \tfrac{1}{2} F$ $=$ $2 R_m$ ⁞ g

$$\tan \tfrac{1}{2} O = \frac{gn}{R_m} \tag{106}$$

Similarly, OPF $= F - O$

Join EF, and in triangle EPF

$$\tan \tfrac{1}{2}(F - O) = \frac{gn}{R_t}$$

$$R_t = \frac{gn}{\tan \tfrac{1}{2}(F - O)} \tag{107}$$

$$\text{chord EF} = E = 2 \left(R_m + \frac{g}{2}\right) \sin \tfrac{1}{2} O \tag{108}$$

Approximate Formulas.

$$D_t = D - D_m \text{ (approx.)} \tag{109}$$

$$E = 2 ng \text{ (approx.)}$$

151. II. When the center of turnout curve lies on the *inside* of main track.

By a process entirely similar it may be shown that

$$\tan \tfrac{1}{2} O = \frac{gn}{R_m} \qquad (110)$$

$$R_t = \frac{gn}{\tan \tfrac{1}{2}(O - F)} \qquad (111)$$

$$E = 2\left(R_m + \frac{g}{2}\right) \sin \tfrac{1}{2} O \qquad (112)$$

Approximate Formulas.

$$D_t = D_m - D \text{ (approx.)} \qquad (113)$$

$$E = 2\, ng \quad \text{(approx.)}$$

152. Example. *Given a 3° curve on main line and a No. 9 frog.*

Required the degree of turnout curve to the inside of the curve.

Table XI, Searles', shows for a No. 9 frog the

degree of curve	$= 7° 31' 04''$; this is ordinarily taken	
	$= 7° 30'$	$= D$
degree of main line	$= 3° 00'$	$= D_m$
degree of turnout	$= 10° 30'$	$= D_t = D + D_m$
By precise formula	$10° 32'$	$= D_t$

The difference between approximate and precise results is commonly small enough to have little effect in the short distance to the point of frog.

In a similar way for a turnout on the *outside* of the main line, using a No. 9 frog, the degree of curve would be

$$D - D_m = 4° 30'$$

Turnouts. 85

153. Split Switch.

The stub switch fails to meet the requirements of modern railroad practice. At the head block, the sliding end of the switch rail is not held firmly down to the ties. Rails are also found to "creep" or travel longitudinally, so that the joint at the head block is often too wide and again is liable to close up so tight that the switch rail cannot be moved.

In turnouts from main line, it is customary now to use the split switch. In this switch the outer rail of the main line and the inner rail of the turnout track are continuous.

The switch rails are steel rails, each planed down at one end to a wedge point, so that it may be caused to lie close against the track rail, and so turn the wheel in the direction intended. An angle is made between the main track and the switch rail. The fixed end of the switch rail is placed at a point corresponding to the head block of the stub switch, and the distance between rails at this point is generally made the same as the "throw" of the stub switch (from 5'' to 6''). The switch rail is often made 15 feet in length, but a length of 19 feet is not uncommon. The "switch angle" is determined by the length of switch rail and this distance between rails (gauge to gauge), and which we may call t.

With the split switch a common practice is to calculate the turnout for a stub switch; the fixed end of the split switch is then placed at the point where the movable end of a stub switch would be placed, and the point of the switch wherever this will bring it. This gives results approximately correct, and sufficiently good to satisfy the requirements of many railroads. For instance if $l = 25$ feet, and the length of switch rail is 15 feet, the *P.C.* of curve will come 10 feet *back from the point of switch*, and the *P.C.* of the center line of turnout curve will be on the center line of main track. The formulas derived for the use of stub switches are available therefore in many cases where split switches are used. *Some prefer a more exact solution.* If the solution following be used, the center line of the turnout curve when produced back until parallel to the main track will not lie on the center line of the main track, but the slight distance from the center may be calculated and an allowance made, so that the details of turnouts to parallel lines can be calculated with simplicity.

154. Problem. *Given, in a turnout, the gauge of track g, length of switch rail l, the distance between rails t, and frog angle F.*
Required the radius of turnout track and distance from movable end of rail to point of frog (or the "lead").

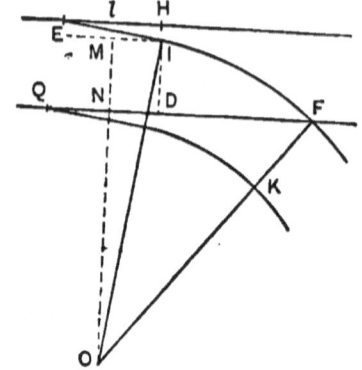

Let EIF and QK be the rails of the turnout.

Draw MI parallel and
OM perpendicular to QF.

Let S = switch angle
t = HI
l = EI = QD

Then $\sin S = \dfrac{t}{l}$

MN   =   MO   −   NO

$$g - t = \left(R + \tfrac{g}{2}\right)\cos S - \left(R + \tfrac{g}{2}\right)\cos F$$

$$R + \tfrac{g}{2} = \frac{g - t}{\cos S - \cos F} \qquad (114)$$

QF =   QD + DF

  = l + $\dfrac{\text{ID}}{\tan \text{IFD}}$

QF = l + $\dfrac{g - t}{\tan \tfrac{1}{2}(F + S)}$ (115)

Some prefer even greater precision, and give weight to the fact that the frog is straight, not curved. It is customary apparently to make the switch rail straight even when used on curves, and it is not customary to use, for curved main line, formulas derived on as strict and perfect a basis as is shown in (114), (115) and in (116) and (117) shown below. It does not appear that the requirements of the work demand such excessive refinement of calculation as would be necessary to do this.

Turnouts.

155. Where the curve continues beyond the point of frog, the introduction of a short tangent (formed by the frog) hurts the alignment so far as appearance is concerned, and also makes less convenient the office work of laying out yards and other turnout work. As a matter of fact, no advantage results so far as the easy running of the train is concerned. The path traveled by the center of a truck is here fixed, not by the line of the outer rail, as is the rule on curves, but by the position of the guard rail near the inside rail, and the actual path is likely to be a double-reversed curve of some sort, being fixed finally, in large part, by the gauge of the inside of the car wheels, and thus varying slightly with different wheels. If it be considered essential to reconcile mathematically the position of the frog (as to length of lead) with the form of the frog, the best way, probably, is to make the frog curved (on both sides) to suit the curvature of the regular turnout curve. The result will be that the frog will truly fit the turnout curve, while on the straight line the frog point will lie outside the proper gauge line by a desirable amount, rendering it possible to make the path of the center of the truck nearly, or quite, a straight line.

156. Problem. *Given, in a turnout, the gauge of track g, length of switch rail l, distance between rails t, length of frog, end to point, k, and the frog angle F.*

Required the radius of turnout curve, and the distance from movable end of rail to point of frog.

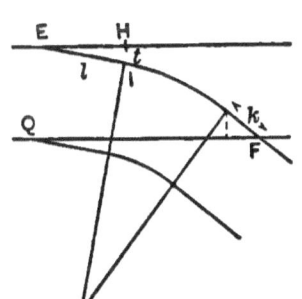

Then following the methods of the preceding problem,

$$\sin S = \frac{t}{l}$$

$$R + \frac{g}{2} = \frac{g - t - k \sin F}{\cos S - \cos F} \quad (116)$$

$$QF = l + \frac{g - t - k \sin F}{\tan \tfrac{1}{2}(F+S)} + k \cos F \quad (117)$$

157. Two parallel straight tracks may be conveniently connected by a turnout in four different ways:

    I. By a reversed curve, the two curves having equal radii.
    II. By a reversed curve, the two curves having unequal radii, and with $P.R.C.$ at point of frog F.
    III. By (*a*) a simple curve to F,
        (*b*) tangent, and return by
        (*c*) simple curve of radius equal to the first.
    IV. By (*a*) a simple curve to F,
        (*b*) tangent, and return by
        (*c*) simple curve of radius unequal to first.

158. I. Problem. *Given the perpendicular distance between two parallel tangents, p; also the common radius, R.*

Required I_r.

Formula (71),   vers $I_r = \dfrac{\frac{1}{2}p}{R}$

159. II. Problem. *Given the radius of the first curve, R_1, also F and p.*

Required the radius of the second curve R_2, to connect the parallel tangents.

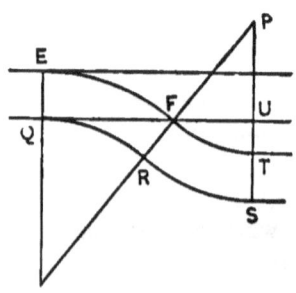

If $P.R.C.$ be taken at F.

Then $I_r = F$
         UT $=$ US $-$ TS
         PT vers TPF $=$ US $-$ TS

$\left(R_2 - \dfrac{g}{2}\right)$ vers $F = p - g$

$R_2 - \dfrac{g}{2} = \dfrac{p - g}{\text{vers } F}$ (118)

160. Problem. *Given, as above, R_1, F, p, n.*

Required R_2.

Second Method.

UT $= p - g$; $R_2 - \dfrac{p}{2} = \dfrac{\text{PU} + \text{PT}}{2}$

by (96) $R_2 - \dfrac{p}{2} = (p - g) 2 n^2$ (119)

161. III. Problem. *Given R, F, p.*

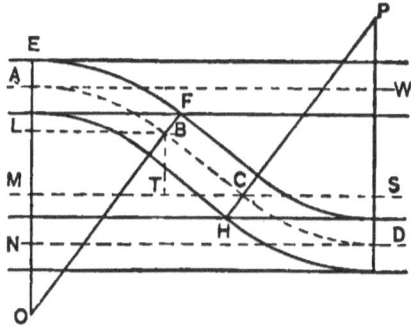

Required the length l of tangent between the two curves of equal radii.

Let AW, ND be the center lines of the parallel tracks, and ABCD the turnout.

Draw the perpendiculars LB, MCS, BT.

Then BT = LM = AN − AL − MN

CB sin BCT = AN − AO vers AOF − PD vers DPC

$l \sin F = p - R \operatorname{vers} F - R \operatorname{vers} F$

$$l = \frac{p - 2R \operatorname{vers} F}{\sin F} \qquad (120)$$

In the case of a cross-over between tracks, it will be convenient to calculate the distance from F to H. Both frog points can then be located and the entire turnout staked out without transit.

162. IV. Problem. *Given R_1, g, p, l, F.*

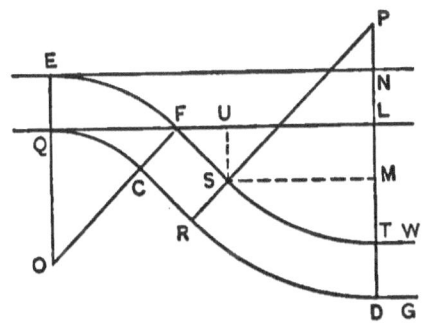

Required R_2.

Let EN and QL, and TW and DG, be the rails of the parallel tangents, and EFST and QCRD the rails of the turnout.

Draw the perpendiculars US, SM.

Then SU = LM = NT − NL − MT

FS sin UFS = NT − NL − PS vers SPM

$l \sin F = p - g - \left(R_2 - \dfrac{g}{2}\right) \operatorname{vers} F$

$$R_2 - \frac{g}{2} = \frac{p - g - l \sin F}{\operatorname{vers} F} \qquad (121)$$

Railroad Curves and Earthwork.

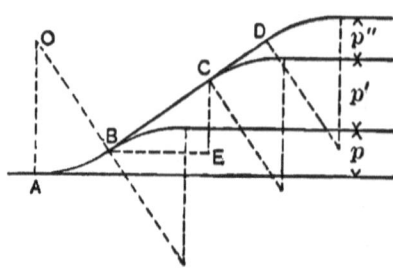

163. Problem. *Given for tracks as shown in figure, the radius R of turnout curve, also the perpendicular distances between tracks p, p', p''.*

Required BC, CD.

From (71) vers $AOB = \dfrac{\frac{1}{2} p}{R}$

$BC \sin CBE = CE$

$BC \sin AOB = p'$

$$BC = \dfrac{p'}{\sin AOB} \qquad (122)$$

Similarly, $CD = \dfrac{p''}{\sin AOB}$

164. Problem. *Given the radial distance p between a given curved main track and a parallel siding, also frog angle F (or number n) and gauge of track g.*

Required the radius of second curve to connect point of frog with siding.

I. When the siding is **outside** the main track.

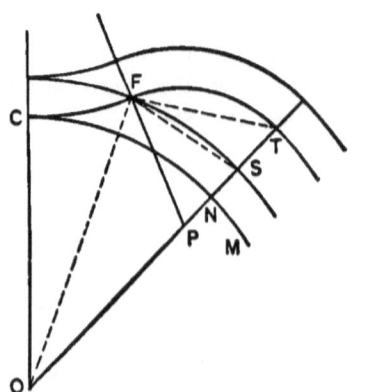

Let CM be the inner rail of the given main line.

CFT inner rail of turnout.

R_m = radius of main line (center).

R_t = radius of turnout (center).

$p = TN$ = radial distance.

Connect FT, FO.

Let $FOT = O$.

Turnouts.

In triangle FTO, $\quad FO = R_m + \dfrac{g}{2}$

$$TO = R_m - \dfrac{g}{2} + p$$

also $\quad$ OFT + OTF = $\quad\quad$ = $180° - O$

$\quad\quad$ OFT − OTF = OFT − PFT = F

Then

$\tan\tfrac{1}{2}$(OFT + OTF) : $\tan\tfrac{1}{2}$(OFT − OTF) = TO + FO : TO − FO

$\cot\tfrac{1}{2} O \quad\quad\quad : \tan\tfrac{1}{2} F \quad\quad\quad = 2R_m + p \; : \; p - g$

$\tan\tfrac{1}{2} O = \dfrac{p-g}{2R_m + p}\cot\tfrac{1}{2} F$

$\quad = \dfrac{p-g}{R_m + \dfrac{p}{2}} \cdot \dfrac{\cot\tfrac{1}{2} F}{2}$

$\tan\tfrac{1}{2} O = \dfrac{(p-g)n}{R_m + \dfrac{p}{2}}$ $\quad\quad\quad\quad\quad\quad$ (123)

Similarly $\quad\quad\quad\quad$ FPT = $F + O$

Join FS.

In the triangle PFS, $\tan\tfrac{1}{2}(F + O) = \dfrac{(p-g)n}{R_t - \dfrac{p}{2}}$;

$$R_t - \dfrac{p}{2} = \dfrac{(p-g)n}{\tan\tfrac{1}{2}(F+O)} \quad\quad (124)$$

Length of curve $\quad\quad\quad L = \dfrac{100(F+O)}{D_t} \quad\quad (125)$

165. Approximate Method.

It might readily be shown that if the entire turnout be calculated as if from a straight track, using the same values of n and p, and the degree of each curve (D_1, D_2) be found; then it would be approximately true that, in the case of a curved main track, the degrees of the turnout curves required would be found by adding or subtracting D_m to or from D_1 and D_2. The distances CF, FT would also be the same as in the turnout from straight track. The demonstrations would follow in principle closely those given in reaching (104), (105).

166. Example.

Turnout from curve *outside* the main track.

Let $D_m = 4$; $n = 9$; $p = 15$; $g = 4.7$.

Precise Method.

$$\tan \tfrac{1}{2} O = \frac{(p-g)n}{R_m + \frac{p}{2}} = \frac{10.3}{1440.2} \times 9 = \frac{92.7}{1440.2}$$

$$R_t - \frac{p}{2} = \frac{(p-g)n}{\tan \tfrac{1}{2}(F+O)}$$

$$= \frac{92.7}{\tan 6°\,51'\,45''}$$

92.7	log 1.967080
1440.2	log 3.158422
$\tfrac{1}{2}O = 3°\,40'\,58''$	tan 8.808658
$\tfrac{1}{2}F = 3°\,10'\,47''$	
$\tfrac{1}{2}(F+O) = 6°\,51'\,45''$	tan 9.080444
92.7	log 1.967080
770.3	log 2.886636
$\tfrac{1}{2}p = 7.5$	
$R_t = 777.8$	
$D_t = 7°\,22'\,17''$	

$$L = \frac{100(F+O)}{D_t} = \frac{100 \times 13°\,43'\,30''}{7°\,22'\,17''} = 186.2$$

167. Approximate Method.

In the case of a turnout from a straight main track, where $n = 9$ and $p = 15$,

$$R_1 - \frac{p}{2} = (p-g)2\,n^2$$

$$= (15.0 - 4.7)2 \times 81 = 1668.6$$

$R = 1676.1$; $D = 3°\,25'$; $F = 6°\,22'$ (Table XI., Searles)

$$L = \frac{100 \times 6°\,22'}{3°\,25'} = 186.3 \text{ for straight tracks}$$

$D_t = D + D_m$

$ = 3°\,25' + 4° = 7°\,25'$ (7° 22' precise method)

$L = 186.3$ as with straight track (186.2 precise method)

Turnouts.

168. II. When the siding is **inside** the main track.

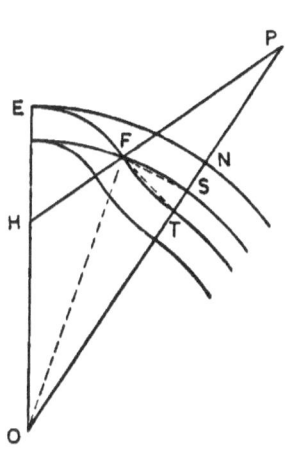

In a similar fashion it may be shown, using this figure, that

From triangle OFT

$$\tan \tfrac{1}{2} O = \frac{(p-g)n}{R_m - \frac{p}{2}} \qquad (126)$$

From triangle PFS

$$R_t - \frac{p}{2} = \frac{(p-g)n}{\tan \tfrac{1}{2}(F-O)} \qquad (127)$$

$$L = \frac{100(F-O)}{D_t} \qquad (128)$$

169. III. When the siding is outside the main track, but with the center of turnout curve inside of main track.

Let EFS be the outer rail of main track.

FT the inner rail of turnout.

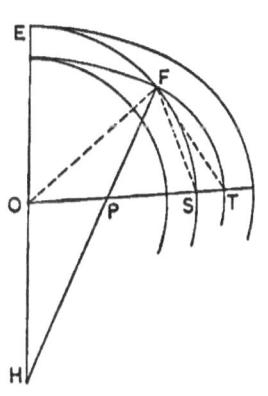

From triangle OFT

$$\tan \tfrac{1}{2} O = \frac{(p-g)n}{R_m + \frac{p}{2}} \qquad (129)$$

From triangle PFS

$$R_t - \frac{p}{2} = \frac{(p-g)n}{\tan \tfrac{1}{2}(F+O)} \qquad (130)$$

$$L = \frac{100(F+O)}{D_t} \qquad (131)$$

With both § 168 and § 169, approximate formulas may be used, the method being similar to that of § 165. Experience will determine in what cases it will be sufficient to use the approximate results, and where precise formulas should be used.

170. Problem. *Given the radial distance between a given curved main track and a parallel siding.*

The two tracks are to be connected by a cross-over, which shall be a reversed curve of given unequal radii.

Required the central angle of each curve of the reversed curve.

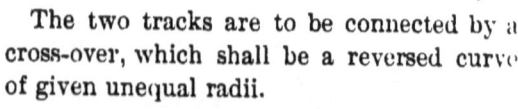

Let ARB be center line of turnout.
AC center line of main track.
AO = R_m
AP = R_1
RQ = R_2

Then in the triangle POQ

PO = $R_m + R_1$
PQ = $R_1 + R_2$
OQ = OC + CB − BQ
 = $R_m + p - R_2$

Solve for OPQ, PQO, POQ, then RQB. In practice this problem might take the following form: *Given* R_m, p, g. Assume n (or F) and n' (or F'). From these calculate R_1 and R_2 (or use Table XI., Searles). Then solve as above.

171. Approximate Method.

Where p is very small compared with R_m, the degree of curve used will frequently be found by the formulas (approx.)

$$D_{t_1} = D - D_m$$
and
$$D_{t_2} = D + D_m.$$

The length of each part may be found for a cross-over between parallel straight tracks, using the same values of n and p, and the same lengths used for this cross-over between curves.

The process is similar in every way to that shown by example in the previous problem.

A similar method of treatment will be applicable in all turnouts from curves where the distance between tracks is not too great.

CHAPTER IX.

"Y" TRACKS AND CROSSINGS.

172. In many cases where a branch leaves a main track, an additional track is laid connecting the two. This is called a "Y" track, and the combination of tracks is called a "Y."

173. Problem. *Given a straight main track, also the P. C. and radius of a simple curve turnout. Also radius of "Y" track.*

Required the distance from P.C. of turnout to P.C. of "Y" track; also the central angles of turnout and of "Y" track to the point of junction.

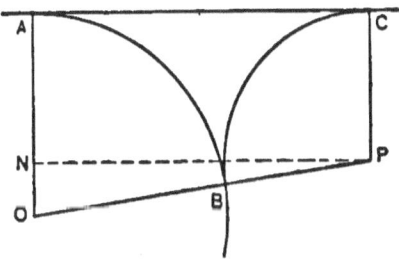

Let AC be the given straight main track.

AB the turnout.

CB the "Y" track.

Draw perpendicular NP.

Let
$$AC = l$$
$$AOB = I_t$$
$$CPB = I_y = 180° - I_t$$

Then
$$\cos AOB = \frac{ON}{OP}$$

$$\cos I_t = \frac{R_t - R_y}{R_t + R_y} \tag{132}$$

$$l = (R_t + R_y) \sin I_t \tag{133}$$

174. Problem. Given a straight main track, also the P.C., radius, and central angle, of simple curve turnout connecting with a second tangent; also the radius of "Y" track.

Required the distance from P.C. of turnout to P.C. of "Y" track, and from P.T. of turnout curve to P.T. of "Y" track.

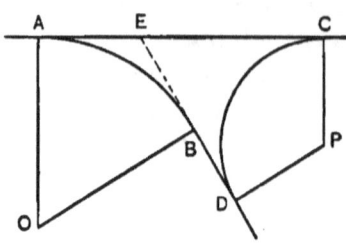

Let AC be the given main track; ABD the turnout; CD the "Y" track.

Let $AO = R_t$; $CP = R_y$
$AC = l$; $BD = m$

Use similar notation for T_t ; T_y ; I_t ; I_y.

Produce BD to E.
Then
$$AC = AE + EC$$
$$= AO \tan \tfrac{1}{2} AOB + CP \tan \tfrac{1}{2} CPD$$
$$l = R_t \tan \tfrac{1}{2} I_t + R_y \cot \tfrac{1}{2} I_t$$
$$l = T_t + T_y \qquad (134)$$
$$BD = ED - EB$$
$$m = T_y - T_t \qquad (135)$$

175. Problem. In the accompanying sketch where
ABC = main track.
AD = turnout.

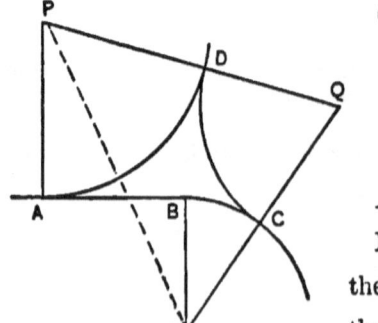

Given $AB = l$
$OB = R_m$
$AP = R_t$
$DQ = R_y$

Required the points D and C.
Find PO, PQ, QO, also OPA
then POQ, OPQ, PQO
then BOC, APQ

D and C will then be easily determined.

"Y" Tracks and Crossings.   97

176. In the figure where ABC is the main track and DC is the turnout.

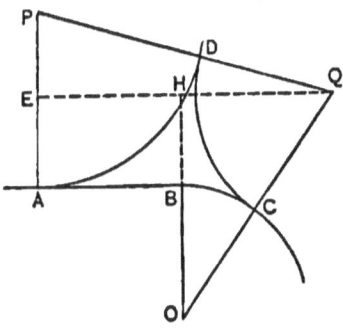

Given $OB = R_m$
$CQ = R_t$
$AP = R_y$
$BOC = O$

Required the points A and D.

Find QH, OH
then   EP
then   EPQ, EQ

then EH = AB,
and   PQO = EQP + OQH
PQO determines position of D
EPQ determines length   AD

177. Problem. *Given a curve crossing a tangent, and the angle C between tangent and curve; also, R, g, g'.*

Required frog angles at A, B, F, D.

Draw AO, BO, CO, FO, DO ; also, MO perpendicular to CM.

Then $MO = R \cos C$

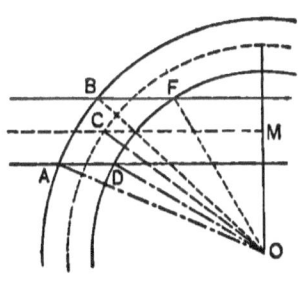

$$\cos A = \frac{MO - \frac{g'}{2}}{R + \frac{g}{2}} \quad (136)$$

$$\cos D = \frac{MO - \frac{g'}{2}}{R - \frac{g}{2}} \quad (137)$$

$$\cos B = \frac{MO + \frac{g'}{2}}{R + \frac{g}{2}} \quad (138)$$

$$\cos F = \frac{MO + \frac{g'}{2}}{R - \frac{g}{2}} \quad (139)$$

178. Problem. *Given radii, R_1, R_2, of two curves crossing at C; also angle at crossing C; also g and g'.*

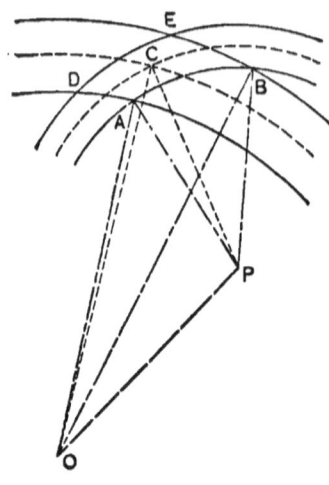

Required, frog angles at A, B, D, E; *also lengths* AB, BE, DE, AD.

Find in triangle OCP, the line OP.

Find in triangle OPA, angles APO, AOP, and OAP $= A$.

Find in triangle OPB, angles BPO, BOP, and OBP $= B$.

Then APB $=$ BPO $-$ APO.

The frog angles at D and E, and the lengths AD, DE, EB, may be calculated in similar fashion.

CHAPTER X.

SPIRAL EASEMENT CURVE.

179. Upon tangent, track ought properly to be level across; upon circular curves, the outer rail should be elevated in accordance with the formula

$$e = \frac{gv^2}{32.2\,R}$$

in which e = elevation in feet

g = gauge of track

v = velocity in feet per second

R = radius of curve in feet.

In passing around a curve, the centrifugal force

$$C = \frac{Wv^2}{32.2\,R}$$

It is desirable for railroad trains that the centrifugal force should be neutralized by an equal and opposite force, and for this purpose, the outer rail of track is elevated above the inner. Any pair of wheels, therefore, rests upon an incline, and the weight W resting on this incline may be resolved into two components, one perpendicular to the incline, the other parallel to the incline, and towards the center of the curve.

The component P parallel to the incline will be

$$P = \frac{We}{g}$$

It will be a very close approximation to assume that C acts parallel to the incline (instead of horizontally). The centrifugal force will be (approximately) balanced if we make

$$P = C$$
$$\frac{We}{g} = \frac{Wv^2}{32.2\,R}$$

or
$$e = \frac{gv^2}{32.2\,R} \qquad (140)$$

In passing directly from tangent to circular curve, there is a point (at $P.C.$) where two requirements conflict; the track cannot be level across and at the same time have the outer rail elevated. It has been the custom to elevate the outer rail on the tangent for perhaps 100 feet back from the $P.C.$ This is unsatisfactory. It is therefore becoming somewhat common to introduce a curve of varying radius, in order to allow the train to pass gradually from the tangent to the circular curve.

180. The transition will be most satisfactorily accomplished when the elevation e increases uniformly with the distance l from the $P.E.$ (point of easement) where the spiral easement curve leaves the tangent; then $\frac{e}{l}$ is a constant, or

$$\frac{gv^2}{32.2\,Rl} = A \text{ (a constant)}$$

$$Rl = \frac{gv^2}{32.2\,A}$$

Since g, v, A, are all constants, we may put the equation in the form
$$Rl = C \qquad (141)$$

181. To further investigate the qualities of this curve, let α = angle of inclination of curve to tangent at any point.

Then
$$R\,d\alpha = dl$$
$$d\alpha = \frac{dl}{R}$$
$$= \frac{l\,dl}{C}$$
$$\alpha = \frac{l^2}{2C} \qquad (142)$$

Again
$$dx = dl \sin \alpha$$
$$dy = dl \cos \alpha$$

and since all values of α within probable limits of spiral easement curves must be small, we may with slight error assume for simplicity that

Spiral Easement Curve.

	$\sin \alpha = \alpha$	
	$\cos \alpha = 1$	
then	$dx = \alpha\, dl$	
	$dy = dl$ (approx.)	
whence	$y = l$	
from (141)	$Ry = C$	(143)
(142)	$\alpha = \dfrac{y^2}{2C}$	(144)

$$dx = \alpha\, dl = \alpha\, dy = \frac{y^2}{2C}\, dy$$

whence integrating $\quad x = \dfrac{y^3}{6C}$ (approx.) $\quad$ (145)

This equation shows that the proper easement curve is. approximately a "Cubic Parabola."

182. The proper easement curve to connect tangent and. curve (or curve and tangent) is, therefore, a curve of constantly and uniformly changing radius. The *Cubic Parabola*, it is seen, is approximately such a curve. *Searles' Railroad Spiral* is also approximately such a curve. Searles changes the radius of his spiral uniformly but not constantly, the change being made at the end of each chord of appreciable length (between 10 and 50 feet each). The resulting curve is so nearly correct that no appreciable error results, whether the Cubic Parabola or Searles' Railroad Spiral be used. The same thing may be said of many other easement curves whose method of application has been set forth by various writers or authors. The error in nearly all cases is absolutely unimportant. Searles' book, The Railroad Spiral, has been so well prepared for convenient use in the field, and presents so great a variety of spirals for selection, that the writer prefers it to any other.

It is desirable, however, to know how to lay out an easement curve without the necessity for using any special book of tables, and the Cubic Parabola described below will be found. simple and available in such a case, though less convenient than the Searles' Spiral when the book of tables of the latter is at hand.

183. Problem. *Given the radius, R, of a Circular Curve. Required to lay out a Cubic Parabola to connect the curve with the tangent.*

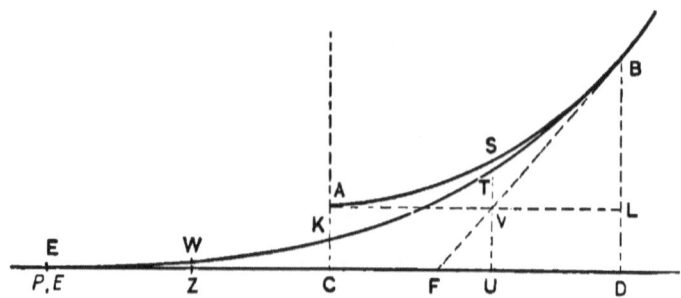

A useful and simple application of the cubic parabola to railroad work is shown by the following:

Field-work.

(a) Lay off tangent ED.

(b) Offset from the proper point C on tangent, to point A from which to run in circular curve AB (offset by any distance AC desired).

(c) Find the distance AL such that BL = 3 AC.

Use (26) $\qquad a = \dfrac{c^2}{2R}$

(d) Make EC = AL.

(e) Mark K so that AK = KC.

(f) Fix any desirable point W by the formula

$$x = x_b \left(\dfrac{y}{y_b}\right)^3$$

(g) Fix any desirable point T by the formula

$$ST = x_b \left(\dfrac{y_b - y}{y_b}\right)^3$$

It will, in general, be sufficiently accurate to take AB = AL and SB = VL.

Spiral Easement Curve. 103

184. It remains to be proved with the field-work indicated above that:

I. The curve passing through the points E, W, K, B, as described, is in fact a cubic parabola.

II. The curves, circular and parabolic, have a common tangent at B.

III. The curves, circular and parabolic, have the same radius at the point B.

IV. The offset ST is correctly calculated.

185. I. In the figure, let

$$AB = \text{the circular curve}$$
$$ED = \text{the tangent}$$
$$EKB = \text{the cubic parabola}$$
$$AL = \text{a tangent to AB at A}$$
$$BF = \text{a tangent to EKB at B}$$

Also
$$BD = 4\,AC$$
$$EC = CD$$
$$AC = LD$$

Then from the equation of the cubic parabola

$$BD = 4\,AC = 8\,KC$$
$$AC = 2\,KC \tag{146}$$
$$BL = BD - LD$$
$$BL = 3\,AC \tag{147}$$

The position of W follows directly from the equation of the cubic parabola.

The ratio $\dfrac{y}{y_b}$ should be simple, for ease in computation.

Let
$$EC = \frac{ED}{2} \quad \text{and} \quad EZ = \frac{EC}{2}$$

then
$$KC = \frac{BD}{8} \quad \text{and} \quad WZ = \frac{KC}{8}$$

186. II. The tangent to the cubic parabola at any point is found thus:

$$\tan \alpha = \frac{dx}{dy} = \frac{\frac{y^2 \, dy}{2\,C}}{dy} = \frac{y^2}{2\,C}$$

For the point B
$$\tan \alpha_b = \frac{y_b^2}{2\,C}$$

but
$$\tan \alpha_b = \frac{BD}{FD} = \frac{x_b}{FD}$$

$$FD = \frac{x_b}{\tan \alpha_b}$$

(145)
$$= \frac{\frac{y_b^3}{6\,C}}{\frac{y_b^2}{2\,C}}$$

$$= \frac{y_b}{3} = \tfrac{2}{3}\,CD$$

By similar triangles $\quad VL = \tfrac{3}{4}\,FD = \tfrac{1}{2}\,CD$

Therefore, for the *cubic parabola* $\quad AV = VL$
and approximately the tangents $\quad AV = VB$

The two tangents to the *circle* at A and B must be equal. Hence when BL is small compared with AL, the tangent to the cubic parabola approximately coincides with that of the circle.

In the case of a very sharp curve, or where BL is large, the approximation may not be sufficiently close.

187. III. From (143) and (145)

At point B
$$R_e = \frac{C}{y_b} = \frac{\frac{y_b^3}{6\,x_b}}{y_b} = \frac{y_b^2}{6\,x_b}$$

$$R_c = \frac{(\text{chord AB})^2}{2\,BL} = \frac{AL^2}{2\,BL} \; (\text{approx.})$$

$$= \left(\frac{y_b}{2}\right)^2 \div \left(2 \times \tfrac{3}{4}\,x_b\right)$$

$$= \frac{y_b^2}{6\,x_b}$$

Therefore, R_c and R_e are approximately equal.

Spiral Easement Curve.

188. IV. *Required in the figure, the offset* ST.

$$US = AC + VS$$

from (26)
$$= \frac{x_b}{4} + BL\left(\frac{AV}{AL}\right)^2 \text{(approx.)}$$

$$= \tfrac{1}{4} x_b + \tfrac{3}{4} x_b \left[\frac{y - \frac{y_b}{2}}{\frac{y_b}{2}}\right]^2$$

$$= \tfrac{1}{4} x_b + \tfrac{3}{4} x_b \left(\frac{4 y^2}{y_b^2} - \frac{4 y}{y_b} + 1\right)$$

$$= \tfrac{1}{4} x_b + \tfrac{3}{4} x_b \left(\frac{4 y^2}{y_b^2} - \frac{4 y}{y_b}\right) + \tfrac{3}{4} x_b$$

$$= x_b + x_b \left(\frac{3 y^2}{y_b^2} - \frac{3 y}{y_b}\right)$$

$$UT = x_b \frac{y^3}{y_b^3}$$

$$ST = US - UT = x_b \left(1 - \frac{3y}{y_b} + \frac{3y^2}{y_b^2} - \frac{y^3}{y_b^3}\right)$$

$$= x_b \left(1 - \frac{y}{y_b}\right)^3$$

$$= x_b \left(\frac{y_b - y}{y_b}\right)^3 = x_b \left(\frac{VL}{ED}\right)^3 \quad (148)$$

If we make VL = EZ, then ST = WZ.

189. Problem. *Given the Radius, R, of a Circular Curve.*

Required to lay out a Cubic Parabola easement curve by the method of Deflection Angles.

Field-work.

(*a*) Lay off the tangent ED.

(*b*) Offset from any point C on tangent, and set point A from which to run in circular curve AB (offset by any distance desired).

(*c*) Find distance AL such that BL = 3 AC.

(*d*) Assuming the arc AB = AL, find the central angle (for the arc AB) = AOB.

(*e*) Take deflection angle DEB = $\dfrac{AOB}{3}$ = i_b.

(*f*) For other deflection angles use the formula $i = i_b \dfrac{y^2}{y_b^2}$

(*g*) With transit at E, run in cubic parabola.

(*h*) With transit at A, run out circular curve beyond B.

190. It remains to be proved that with this field-work

A. The formula $i = i_b \dfrac{y^2}{y_b^2}$ is correct.

B. The deflection angle DEB = $\dfrac{AOB}{3}$

A. Let i = deflection angle from $P.E.$ to any point xy.

Since
$$x = \frac{y^3}{6\,C}$$

$$\tan i = \frac{x}{y} = \frac{y^2}{6\,C}$$

For small angles, tangents are proportional to the angles. Here i is always small.

Therefore $\qquad i : i' = y^2 : y'^2$ (approx.) $\hfill$ (149)

191. B. In curve AB.

$$BL = \frac{AB^2}{2R} \quad \text{(AB here being the chord)}$$

$$BD = \frac{4}{3} \cdot \frac{AB^2}{2R} = \frac{2\,AB^2}{3R}$$

$$ED = 2\,AB \quad \text{(approx.)}$$

$$\tan i_b = \frac{BD}{ED} = \frac{AB}{3R}$$

$$\tan \tfrac{1}{2} AOB = \frac{BL}{AL} = \frac{AB^2}{AL \times 2R} = \frac{AB}{2R} \quad \text{(approx.)}$$

Tangents of small angles are proportional to the angles.

Hence
$$i_b : \frac{AOB}{2} = 2 : 3$$

$$i_b = \frac{AOB}{3} \tag{150}$$

192. For most easement curves likely to be used, the approximations here given will lead to no errors of consequence. The greatest lack of precision will be in the position of the point B, and since it is undesirable that any error, however small, should be carried ahead, if it be possible to avoid it, there is an advantage in running in the circular curve with the transit at A, and with a back-sight parallel to the tangent ED, rather than with the transit at B.

If preferred, the length of line AL or AB may be assumed and AC calculated. This method will simplify the calculation of angles if AL or AB be taken in round numbers.

193. Compound Curves. In the case of Compound Curves, it is proper and desirable that easement curves should be introduced between the two circular curves forming the compound curve.

Problem. *Given the Degrees D_l, D_s, of a Compound Curve, and the offset between them.*

Required to connect the circular curves by a Cubic Parabola.

Calculate a cubic parabola suitable to connect a tangent and a circular curve whose degree is $D_l - D_s$, and for the given offset p (= AC, p. 102).

The ordinates calculated may then be laid off from either circular curve as is most convenient. The middle point of the cubic parabola will be on the offset measured by p and midway between the two circular curves. The principle involved is the same as that discussed in connection with turnouts from curves, where the approximation was pointed out (pages 84, 92, 94).

194. Problem. *Given in a Compound Curve, D_l, D_s, p.*

Required the Deflection Angles for a Cubic Parabola to connect the circular curves.

(a) Find by § **189** the Deflection Angles proper for a Cubic Parabola to connect a tangent with a circular curve of degree $= D_l - D_s$.

Let these $= i_1$, i_2, i_3, etc.

(b) Find the deflection angles to corresponding points on one of the circular curves, the auxiliary tangent for these being at the point where the cubic parabola leaves this circular curve (where the transit will be set).

Let these $= \dfrac{d_1}{2}, \dfrac{d_2}{2}, \dfrac{d_3}{2}$, etc.

(c) The total deflections required will be for

point 1 $\dfrac{d_1}{2} + i_1$

point 2 $\dfrac{d_2}{2} + i_2$

point 3 $\dfrac{d_3}{2} + i_3$ etc.

Spiral Easement Curve.

195. Problem. *Given the degrees of two parts of a compound curve.*

Required to connect the two curves by a Searles' Spiral; and to find the offset, and also the points where the spiral leaves one curve and again joins the other.

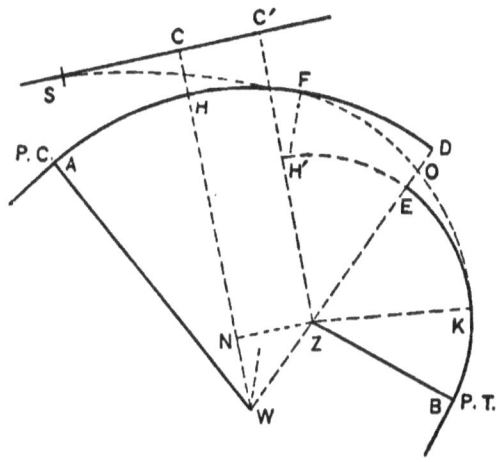

Let the curves be

AD with center W
H'EB " " Z

The required spiral FK must be selected so that at F it will have the proper curvature to join the circle W, and at K the proper curvature for circle Z. If the spiral be continued *back* to the *P.S.* at S, we may show the tangent at S by SCC'. The points F and K must be regular points on the spiral.

Draw perpendiculars WC and ZC' and ZN, and extend circle Z to H'.

Then, using Searles' notation, $SC = q_l$; $CH = p_l$
$SC' = q_s$; $C'H' = p_s$

The offset DE is required.

Having selected a spiral which shall properly fit at F and K, find for this spiral q_l, p_l, q_s, p_s.

Then, $$\tan CWD = \frac{ZN}{NW} = \frac{q_s - q_l}{R_l + p_l - (R_s + p_s)} \qquad (151)$$

$$DE = DW - ZE - ZW$$
$$= R_l - R_s - \frac{q_s - q_l}{\sin CWD} \qquad (152)$$

The point F may be found from D since angle
$$FWD = CWD - s_f.$$

The point K may be found by similar process.

The point D is assumed on the ground and is the point selected for compounding the curves.

196. Example. *Given a compound curve, 2° and 5°.*

Required to select a spiral (Searles); to find offset between 2° and 5° curves; also points where the spiral connects with 2° and 5° curves.

From Table VI., Searles' Spiral,
and " IV., " Field Engineering,
select a Spiral and complete the following Table:

Degree	$n-c$	g	p	R	$R+p$
2°	3 — 33	48.997	0.236	2864.93	2865.17
5°	9 — 33	146.846	3.855	1146.28	1150.14

Difference, 97.849   1718.65   1715.03

$$97.849 \quad \log 1.990556$$
$$1715.03 \quad \log 3.234272$$

3° 15′ 55″ tan 8.756284   sin 8.755579
            97.849   log 1.990556

           1717.82   log 3.234977
$R_l - R_s =$ 1718.65
  offset   =   0.83

Find point F.

$$s_2 = \frac{\begin{array}{r}3° \ 15′ \ 55″\\ 1°\end{array}}{2° \ 15′ \ 55″} = 2°.159167′ = 2.2653°(2° c$$
$$113.26 = FD$$

Find point K.

$$s_5 = \frac{\begin{array}{r}7° \ 30′\\ 3° \ 15′ \ 55″\end{array}}{4° \ 14′ \ 05″} = 4° \ 14087′ = 4.2348°(5° c$$
$$84.70 = EK$$

Length spiral $n - c$
    6 — 33 = 198.00   197.96 = FD + EK
                                  check.

CHAPTER XI.

SETTING STAKES FOR EARTHWORK.

197. The first step in connection with Earthwork is staking out, or "**Setting Slope Stakes**," as it is commonly called.

There are two important parts of the work of setting slope stakes:

                I. Setting the stakes.

              II. Keeping the notes.

The data for setting the stakes are:

(*a*) The ground with center stakes set at every station (sometimes oftener).

(*b*) A record of bench marks, and of elevations and rates of grades established.

(*c*) The base and side slopes of the cross-section for each class of material.

In practice, notes of alignment, a full profile, and various convenient data are commonly given in addition to the above.

198. I. Setting the Stakes. The work consists of:

(*a*) Marking upon the back of the center stakes the "cut" or "fill" in feet and tenths, as

            C 2.3   or   F 4.7.

(*b*) Setting side stakes or slope stakes at each side of the center line at the point where the side slope intersects the surface of the ground, and marking upon the inner side of the stake the "cut" or "fill" at that point.

112 Railroad Curves and Earthwork.

199. (a) The process of finding the cut or fill at the center stake is as follows:

Given for any station the height of instrument $= h_i$, *and the elevation of grade* $= h_g$.

Then the required *rod reading for grade*

$$r_g = h_i - h_g. \tag{153}$$

It is not necessary to figure h_g for each station.

Let
$$h_{g_0} = h_g \text{ at Sta. } 0$$
$$h_{g_1} = h_g \text{ " " } 1$$
$$h_{g_2} = h_g \text{ " " } 2, \text{ etc.}$$

Also use similar notation for r_g.

Let $\quad g =$ rate of grade (rise per station)

Then
$$h_{g_1} = h_{g_0} + g$$
$$h_{g_2} = h_{g_1} + g$$
$$h_{g_3} = h_{g_2} + g, \text{ etc.}$$

$$r_{g_0} = h_i - h_{g_0}$$
$$r_{g_1} = h_i - h_{g_1}$$
$$\quad = h_i - (h_{g_0} + g) = h_i - h_{g_0} - g$$
$$r_{g_1} = r_{g_0} - g \tag{154}$$

Similarly, $\quad r_{g_2} = r_{g_1} - g$, etc.

It will be necessary, or certainly desirable, to figure h_g and r_g anew for each new h_i. It is well to figure h_g and r_g (as a check) for the last station before each turning point.

Setting Stakes for Earthwork. 113

200. Example. $h_i = 106.25$

Sta. 0, grade elevation 100.00 rate $+1.00$
   5,   "   "   105.00   " $+0.50$
  10,   "   "   107.50

$r_{g_0} = 106.25 - 100.00 = 6.25$ 6.25

$r_{g_1} =$ $6.25 - \overset{g}{1.00} \Big\} = 5.25$

$r_{g_2} =$ $5.25 - 1.00 = 4.25$

$r_{g_3} =$ $4.25 - 1.00 = 3.25$

$r_{g_4} =$ $3.25 - 1.00 = 2.25$

$r_{g_5} =$ $2.25 - 1.00 = 1.25$

Change in rate

$r_{g_6} =$ $1.25 - 0.50 = 0.75$

$r_{g_7} =$ $0.75 - 0.50 = 0.25$

It is found necessary to take a *T.P.* here, and we therefore find

$h_{g_7} = h_{g_5} + 2g$
$\phantom{h_{g_7}} = 105.00 + 1.00 = 106.00$

$r_{g_7} = h_i - h_{g_7} = 106.25 - 106.00 = 0.25$

Therefore all intermediate values r_{g_1}, r_{g_2}, etc., are "checked."

201. Having thus found r_g, next, by holding the rod upon the surface of the ground at the center stake, the rod reading $r_c = LO$ is observed from the instrument. The cut or fill

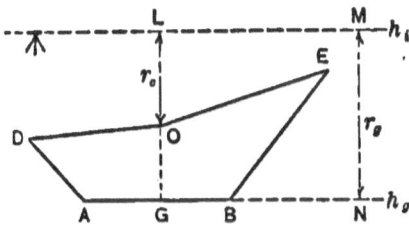

$c = OG = MN - LO$
$ = r_g - r_c$ (155)

In the figure given the values of r_g and c are positive; a positive value of c indicates a "cut," a negative value of c indicates a "fill."

It can be shown that in the two cases of "fill,"

(1) When h_i is greater than h_g, and
(2) When h_i is less than h_g,

the formula given will hold good by paying due attention to the sign of r_g, whether $+$ or $-$.

202. (b) **Setting the Stake for the Side Slope.**

(1) *When the surface is level.*

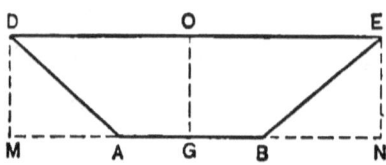

Let $b = \text{AB} = $ base of section

$c = \text{OG} = $ center height

$s = \dfrac{\text{BN}}{\text{EN}} = \dfrac{\text{AM}}{\text{DM}} = $ side slope

$d = \text{OD} = \text{OE} = $ distance out

Then $d = \text{GB} + \text{BN}$

$= \tfrac{1}{2}b + s \times \text{DM} = \tfrac{1}{2}b + s \times \text{EN}$

$= \tfrac{1}{2}b + sc$

203. Setting the Stake for the Side Slope.

(2) *When the surface is not level.*

Here the process is less simple.

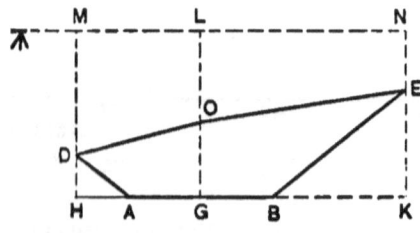

Let $b = \text{AB} = $ base

$c = \text{OG} = $ center height (or cut)

$s = $ slope

Setting Stakes for Earthwork. 115

h_r = EK = side height right

h_l = DH = "   " left

d_r = GK = distance out right

d_l = GH =   "   " left

Then
$$\left.\begin{array}{l}d_r = \tfrac{1}{2}b + sh_r \\ d_l = \tfrac{1}{2}b + sh_l\end{array}\right\} \quad (156)$$

But h_r and h_l are not known. It is evident from the figure that $h_r > c$ and $h_l < c$ in the case indicated, and therefore

$$d_r > \tfrac{1}{2}b + sc$$
$$d_l < \tfrac{1}{2}b + sc$$

204. It would be possible in many cases to take measurements such that the rate of slope of the lines OE and OD would be known, and the positions of E and D determined by calculation from such data. But speed and results finally correct are the essentials in this work, and these are best secured by finding h_l and h_r and the corresponding d_l and d_r upon the ground by a series of approximations, as described below.

Having determined c, use this as a basis, and make an estimate at once as to the probable value of h_r at the point where the side slope will intersect the surface, and calculate $d_r = \tfrac{1}{2}b + sh_r$ to correspond.

Measure out this distance, set the rod at the point thus found, take the rod reading on the surface, and if the cut or fill thus found from the rod reading yields a value of d_r equal to that actually measured out, the point is correct. Otherwise make a new and close approximation from the better data just obtained, always starting with h_r and calculating d_r, and repeat the process until a point is reached where the cut or fill found from the rod reading yields a distance out equal to that taken on the ground. Then set the stake, and mark the cut or fill corresponding to h_r upon the inner side, as previously stated.

Perform the same operation in a similar way to determine $d_l = \tfrac{1}{2}b + sh_l$, and mark this stake also upon the inner side with a cut or fill equal to h_l.

205. It requires a certain amount of work in the field to appreciate the process here outlined, but which in practice is very simple. It may impress some as being unscientific, and at first trial as slow, but with a little practice it is surprising how rapidly, almost by instinct, the proper point is reached, often within the required limits of precision at the first trial, while more than two trials will seldom be necessary, except in difficult country.

206. The instrumental work is the same in principle as at the center stake.

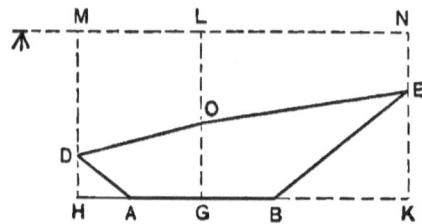

Let $r_r = \mathsf{NE} =$ rod reading at slope stake right,

then $\mathsf{KN} - \mathsf{NE} = r_g - r_r = h_r$

here r_g is the same for center, right and left of section.

In some cases it may be necessary to make one or more resettings of the level in order to reach the side stakes from the center stake. In this case, of course, a new r_g must be calculated from the new h_i. This introduces no new principle, but makes the work slower.

207. A "slope-board" or "level-board" may be used to advantage in many cases. In certain sections of country this might be considered almost indispensable. It consists simply of a long, straight-edge of wood (perhaps 15 ft. long) with a level mounted in the upper side. It is used with any self-reading rod. A rod quickly hand marked will serve the purpose well. Having given the cut or fill at the center, or at any point in the section, the leveling for the side stakes, and for any additional points, can readily, and with sufficient accuracy, be done by this "level-board," and the necessity for taking new turning points and resetting the level avoided.

Setting Stakes for Earthwork. 117

208. II. *Keeping the Notes.*

The form of note-book used for keeping the notes of slope stakes and of center cuts and fills, often called "cross-section" notes, is shown on the following two pages.

The left-hand column for stations should read from bottom to top.

The surface elevations in column 2 are *not obtained directly from the levels*, but result from adding to the grade elevation at any station the cut or fill at that station, paying due attention to the signs. This column of surface elevations need not be entered up in the field, but may be filled in as office work more economically.

The column of grade elevations consists of the grade elevations as figured for each station.

The figures marked + are cuts in feet and tenths, and those marked − are fills; the figures above the cuts and fills are the distances out from the center, and the position in the notes, whether right or left of the center, corresponds to that on the ground.

The columns on the right-hand page are used for entering, when computed, the "quantities," or number of cubic yards, in each section of earthwork.

209. The column "**General Notes**" is used for entering extra measurements (of ditches, etc.) not included in the regular cross-section notes; also notes of material "hauled"; classification of material and various other matters naturally classed under the head of "Remarks."

210. When the surface is irregular between the center and side stakes, additional rod readings and distances out are taken, and the results entered as shown for station 0 on p. 118, the section itself being as shown below in the sketch.

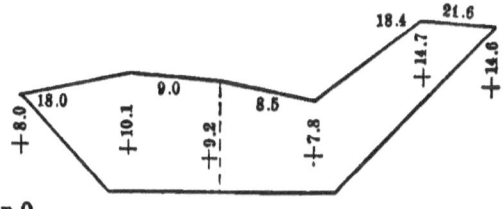

Station 0

118 Railroad Curves and Earthwork.

211. Form of Cross-Section Book (left-hand page).

(Date)
(Names of Party)

Station	Surface Elev.	Grade Elev.	Cross-Section				
5	97.1	105.00	$\frac{18.4}{-7.6}$	-7.9	$\frac{19.4}{-8.3}$		
+69.7 P.T.	94.4	104.70	$\frac{22.1}{-10.1}$	-10.3	$\frac{23.0}{-10.7}$		
4	96.9	104.00	$\frac{19.3}{-8.2}$	-7.1	$\frac{17.0}{-6.7}$		
+27.2 P.C.	98.0	103.27	$\frac{16.6}{-6.4}$	-5.8	$\frac{12.4}{-3.6}$		
3	98.1	103.00	$\frac{16.0}{-6.0}$	-4.9	$\frac{10.9}{-2.6}$		
+91	100.6	102.91	$\frac{13.3}{-4.2}$	-2.3	$\frac{10.0}{0.0}$		
+76	102.8	102.76	$\frac{10.3}{-2.2}$	0.0	$\frac{11.9}{+1.9}$		
+64	103.7	102.64	$\frac{10.0}{0.0}$	$+1.1$	$\frac{13.2}{+3.2}$		
+50	106.4	102.50	$\frac{13.4}{+3.4}$	$+3.9$	$\frac{17.1}{+7.1}$		
2	115.1	102.00	$\frac{16.7}{+6.7}$	$+13.1$	$\frac{26.7}{+16.7}$		
1	117.7	101.00	$\frac{22.7}{+12.7}$	$\frac{10.0}{+17.2}$	$+16.7$	$\frac{10.0}{+13.1}$	$\frac{22.2}{+12.2}$
0	109.2	100.00	$\frac{18.0}{+8.0}$	$\frac{9.0}{+10.1}$	$+9.2$	$\frac{8.5}{+7.8}\ \frac{18.4}{+14.7}\ \frac{24.6}{+14.6}$	

Setting Stakes for Earthwork.

212. (Right-hand Page.)

	Excavation		Embank-ment	General Notes
L. Rock	S. Rock	Earth		

213. Cross-sections are taken at every full station, at every *P.C.* or *P.T.* of curve, wherever grade cuts the surface, and in addition, at every break in the surface. In the figure below, showing a profile, sections should be taken at the following stations: —

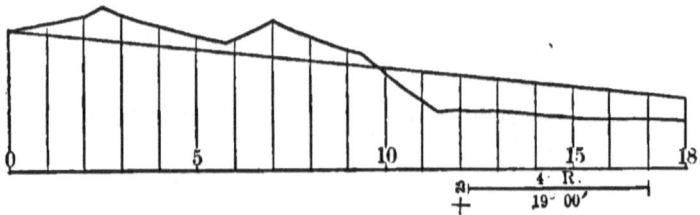

At Stations 0, 1, 2, 2 + 52, 3, 4, 5, 5 + 80, 6, 7, 8, 9, 9 + 29, 9 + 82, 10, 11, 11 + 30, 12, 12 + 25 *P.C.*, 13, 14, 15, 16, 17 *P.T.*, 18.

214. It is not necessary actually to drive stakes in *all* cases where a cross-section is taken and recorded, but in every case where they will aid materially in construction stakes should be set. It is best to err on the safe side, which is the liberal side. In passing from cut to fill, it is customary to take full cross-sections, not only at the point where the grade line cuts the surface at the *center* line of survey, but also where the grade cuts the surface at the outside of the base, both *right* and *left*, as in the figure below, which illustrates the notes on p. 118; full cross-sections are taken not only at stations 2 + 76, but also at 2 + 64 and 2 + 91.

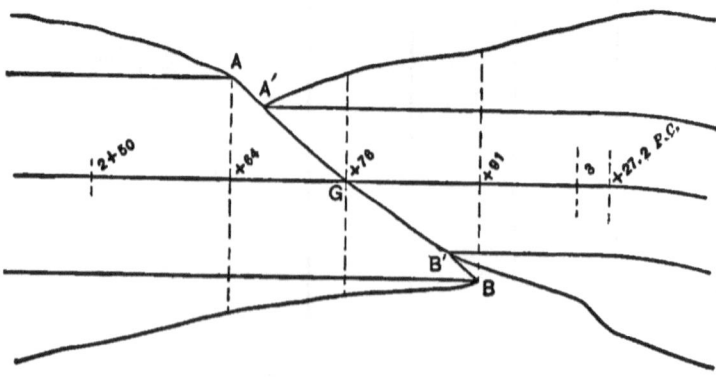

Setting Stakes for Earthwork.

215. Stakes are actually set at the center G and at the points A and B, where the outside line of the base of *Excavation* cuts the surface. It is not customary to set stakes or record the notes for the points A' and B', where the outside line of the base of *Embankment* cuts the surface. The stakes at A, G, and B are a sufficient guide for construction, and the solidities or "quantities" would in general be affected only slightly by the additional notes if they were made. When the line AGB crosses the center line nearly at right angles, it would not be necessary to take more than one section so far as the notes are concerned. It is well, however, to set the stakes A and B exactly in their proper position.

216. Wherever an opening is to be left in an Embankment for a bridge or for any other structure, stakes should be set as in the figure below: —

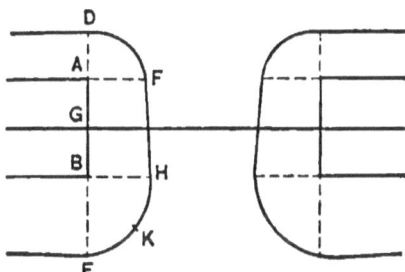

At A and B (at the side of the base and top of the slopes AF and BH) stakes should be set marked "*Bank to Grade*"; and at F and H (at the foot of the slopes) stakes should be set marked "*Toe of Slope.*" Where the bank is high, an additional stake K at foot of slope may be set as an aid to construction. The stakes at D and E should also be set as ordinary slope stakes.

217. The "level notes" proper, or the record of heights of instrument, bench marks, turning points, etc., used in setting slope stakes, are usually kept separate from the cross-section notes. One reason for this is that level notes run from top to bottom of page, while cross-section notes read from bottom to top of page. The level notes should be kept either in the back of the cross-section book or in a level book carried for that purpose. Keeping these or any other notes on a slip of paper is bad practice.

218. Earthwork can be most readily computed when the section is a "*Level Section*," that is when the surface is level across the section; but this is seldom the case, and for purposes of final computation it is not often attempted to take measurements upon that basis.

219. In general, in railroad work, the ground is sufficiently regular to allow of "*Three-Level Sections*" being taken, one level (elevation) at the center and one at each slope stake, as shown by these notes, where Base is 20, and Slope $\frac{1}{2}$ to 1: —

$$\frac{11.3}{+2.6} \qquad +4.2 \qquad \frac{12.8}{+5.5}$$

The term "*Three-Level Section*" is usually applied only to *regular* sections where the widths of base in each side of the center are the same. In regular three-level sections the calculation of quantities can be made quite simple. To facilitate the final estimation of quantities, it is best to use three-level sections as far as possible.

220. In many cases where three-level sections are not sufficient, it may be possible to use "*Five-Level Sections*," consisting of a level at the center, one at each side where the *base* meets the side slope, and one at each side slope stake, as shown by the following notes: —
Base 20, Slope 1 to 1,

$$\frac{22.7}{+12.7} \qquad \frac{10.0}{+17.2} \qquad +16.7 \qquad \frac{10.0}{+13.1} \qquad \frac{22.2}{+12.2}$$

The term "*Five-Level Section*" is usually applied only to regular sections where the base and the side slopes are the same on each side of the center.

221. Where the ground is very rough, levels have to be taken wherever the ground requires, and the calculations must be made to suit the requirements of each special case, although certain systematic methods are generally applicable. Such sections are called "*Irregular Sections*."

CHAPTER XII.

METHODS OF COMPUTING EARTHWORK.

222. In calculating the solidities or "quantities" of Earthwork, the principal methods used are as follows : —

    I. AVERAGING END AREAS.

    II. PRISMOIDAL FORMULA.

    III. MIDDLE AREAS.

    IV. EQUIVALENT LEVEL SECTIONS.

    V. MEAN PROPORTIONALS.

    VI. HENCK'S METHOD.

223. I. Averaging End Areas.

This is the simplest method : —

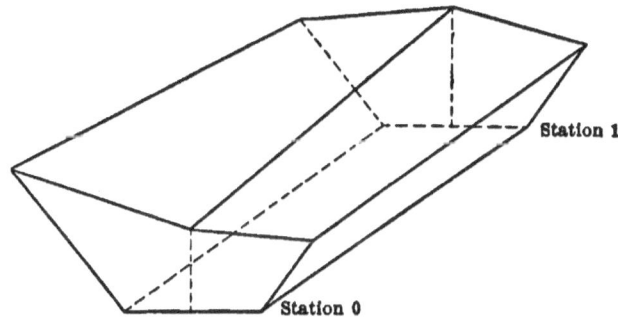

Let A_0 = area of cross-section at Station 0

    A_1 = " " " " " " 1

    l = length of section, Sta. 0 to Sta. 1

    S = solidity of section of earthwork (Sta. 0 to 1)

124   *Railroad Curves and Earthwork.*

Then $$S = \frac{A_0 + A_1}{2} l \text{ (in cubic feet)} \tag{158}$$

$$= \frac{A_0 + A_1}{2} \cdot \frac{l}{27} \text{ (in cubic yards)} \tag{159}$$

As (158) is capable of expression

$$S = A_0 \frac{l}{2} + A_1 \frac{l}{2}$$

it is practically based on the assumption that the solidity consists of two prisms, one of base A_0 and one of base A_1, and each of a length, or altitude of $\frac{l}{2}$.

224. To use this method, we must find the area A of each cross-section; the cross-section may be: —

(a) *Level.*

(b) *Three-Level.*

(c) *Five-Level.*

(d) *Irregular.*

225. (a) **Level Cross-Section.**

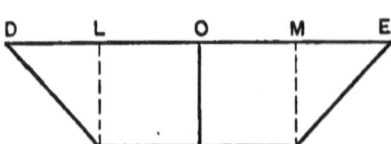

Let b = base = AB

s = side slope = $\dfrac{DL}{AL} = \dfrac{EM}{BM}$

c = center ht. = OG

A = area of cross-section

Methods of Computing Earthwork.

Then DL = EM = sc

$$A = AB' \times OG + DL \times AL$$
$$= bc + sc^2$$
$$= c(b + sc) \qquad (160)$$

226. (b) **Three-Level Section.** First Method.

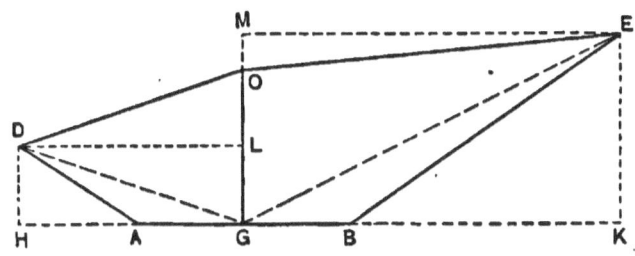

Let b = base = AB

s = side slope

c = center ht.

h_r = side height EK

h_l = "   "   DH

d_r = distance out ME

d_l = "   "   DL

A = area of cross-section

Then A = OGD + OGE + GBE + AGD

$= \frac{1}{2} OG \times DL + \frac{1}{2} OG \times ME + \frac{1}{2} GB \times EK + \frac{1}{2} AG \times DH$

$$= \tfrac{1}{2} c(d_l + d_r) + \tfrac{1}{2} \frac{b}{2}(h_r + h_l)$$

$$= \frac{c(d_l + d_r) + \dfrac{b}{2}(h_l + h_r)}{2} \qquad (161)$$

227. (*b*) **Three-Level Section.** Second Method.

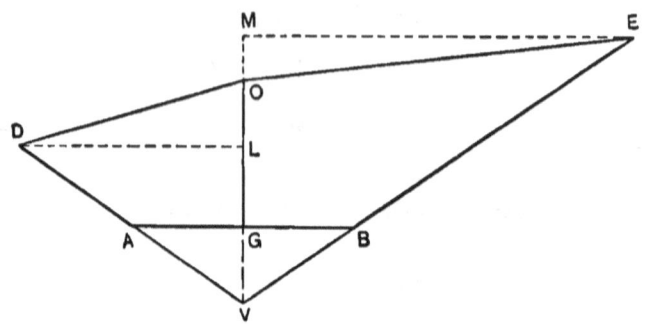

Using the same notation.

$$\frac{GB}{GV} = s$$

$$GV = \frac{GB}{s} = \frac{b}{2s}$$

$$OV = c + GV = c + \frac{b}{2s}$$

The triangle ABV is often called the "Grade Triangle."

$$\text{Area ABV} = GV \times GB$$

$$= \frac{b^2}{4s}$$

$$\text{Area EODV} = OV \times \frac{DL}{2} + OV \times \frac{ME}{2}$$

$$= \left(c + \frac{b}{2s}\right)\frac{d_l + d_r}{2}$$

$$A = \text{EODV} - \text{ABV}$$

$$= \left(c + \frac{b}{2s}\right)\frac{d_l + d_r}{2} - \frac{b^2}{4s}$$

Let
$$D = d_l + d_r$$

$$A = \left(c + \frac{b}{2s}\right)\frac{D}{2} - \frac{b^2}{4s} \qquad (162)$$

In using this formula for a *series* of cross-sections of the same base and slope, $\dfrac{b}{2s}$ and $\dfrac{b^2}{4s}$ are constants, and the computation of A becomes simple and more rapid than the first method.

Methods of Computing Earthwork. 127

228. (c) **Five-Level Section.**

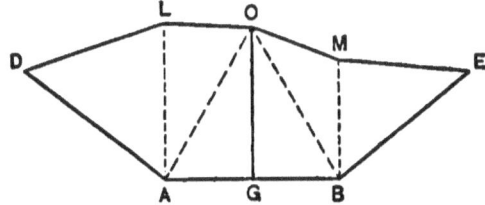

Use notation the same as before; in addition let

$$f_r = \text{height MB}$$
$$f_l = \quad`` \quad \text{LA}$$

Then
$$A = \text{AOB} + \text{DLOA} + \text{EMOB}$$
$$= \frac{cb}{2} + \frac{f_r d_r}{2} + \frac{f_l d_l}{2}$$
$$A = \frac{cb + f_r d_r + f_l d_l}{2} \tag{163}$$

229. (d) **Irregular Section.**

The "Irregular Section," as shown in the figure, may be divided into trapezoids by vertical lines, as in Fig. 1; or into triangles by vertical and diagonal lines, as in Fig. 2.

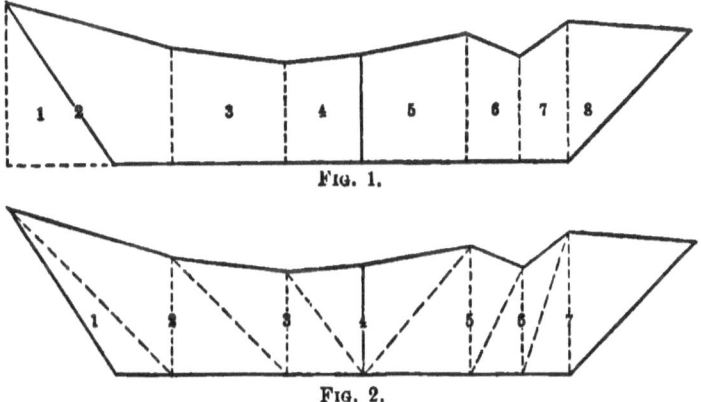

Fig. 1.

Fig. 2.

The triangles in Fig. 2 can be computed in groups of two each pair having a common base (vertical).

It will be seen that Fig. 1 requires 8 solutions and Fig. 2 only 7 solutions of trapezoids or triangles. The computations can be made with substantially equal simplicity in either case, and after a little experience, directly from the notes without any necessity for a sketch.

230. Another method which has been used for calculating irregular cross-sections is to plat them on cross-section paper, and get the area by "*Planimeter*." In very irregular cross-sections this method would prove economical as compared with direct computation by ordinary methods, but it is probable that in almost every case equal speed and equal precision can be obtained by the use of suitable tables or diagrams (to be explained later) ; for this reason the use of the planimeter is not recommended.

231. Having found the values of A for each cross-section, S is found in each case by the formula above given,

$$S = \frac{A_0 + A_1}{2} \cdot \frac{l}{27} \text{ (in cu. yds.)} \qquad (159)$$

It is found that this formula is only approximately correct. Its simplicity and *substantial accuracy* in the majority of cases render it so valuable that it has become the formula in most common use. It gives results, in general, larger than the true solidity.

232. II. Prismoidal Formula.

"A prismoid is a solid having for its two ends any dissimilar parallel plane figures of the same number of sides, and all the sides of the solid plane figures also."

Any prismoid may be resolved into prisms, pyramids, and wedges, having as a common altitude the perpendicular distance between the two parallel end planes.

Let A_0 and A_1 = areas of end planes.

M = area of middle section parallel to the end planes.

l = length of prismoid, or perpendicular distance between end planes.

S = solidity of the prismoid.

Then it may be shown that

$$S = (A_0 + 4M + A_1)\frac{l}{6}$$

Methods of Computing Earthwork.

233. Let B = area of lower face, or base of a prism, wedge, or pyramid.

b = area of upper face.

m = middle area parallel to upper and lower faces.

a = altitude of prism, wedge, or pyramid.

s = solidity "   "   "   "   "

Then the area of the *upper face* b in terms of *lower base* B will be for

Prism	Wedge	Pyramid
$b = B$	$b = 0$	$b = 0$

and the *middle area* m will be for

Prism	Wedge	Pyramid
$m = B$	$m = \dfrac{B}{2}$	$m = \dfrac{B}{4}$

The solidity s will be for

Prism

$$s = aB = \frac{a}{6} \cdot 6B = \frac{a}{6}(B + 4B + B) = \frac{a}{6}(B + 4m + b)$$

Wedge

$$s = \frac{aB}{2} = \frac{a}{6} \cdot 3B = \frac{a}{6}\left(B + \frac{4B}{2} + 0\right) = \frac{a}{6}(B + 4m + b)$$

Pyramid

$$s = \frac{aB}{3} = \frac{a}{6} \cdot 2B = \frac{a}{6}\left(B + \frac{4B}{4} + 0\right) = \frac{a}{6}(B + 4m + b)$$

Since a prismoid is composed of prisms, wedges, and pyramids, the same expression may apply to the prismoid, and this may be put in the general form

$$S = (A_0 + 4M + A_1)\frac{l}{6} \qquad (163)$$

using the notation of the preceding page.

234. A regular section of earthwork having for its surface a plane face is a prismoid. Most sections of earthwork have not their surface plane, and are not strictly prismoids, although they are so regarded by some writers.

In this figure the lines E_0O_0 and E_1O_1 are not parallel, and therefore the surface $O_0O_1E_1E_0$ is not a plane. The most common assumption as to this surface is that the lines O_0O_1 and E_0E_1 are right lines, and that the surface $O_0O_1E_1E_0$ is a warped surface, generated by a right line moving as a generatrix always

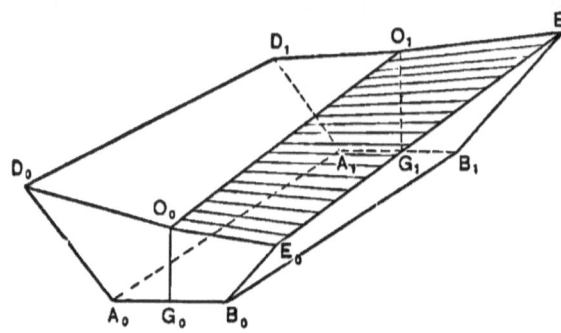

parallel to the plane $O_0G_0B_0E_0$ and upon the lines O_0O_1 and E_0E_1 as directrices, as indicated in the figure. The surface thus generated is a warped surface called a "hyperbolic paraboloid." It will be shown that the "prismoidal formula" applies also to this solid, which is not, however, properly a prismoid.

235. In the following figure, which has perpendicular sides $D_0A_0A_1D_1$, $E_0B_0B_1F$ and the lines D_0E_0 and D_1E_1 right lines,

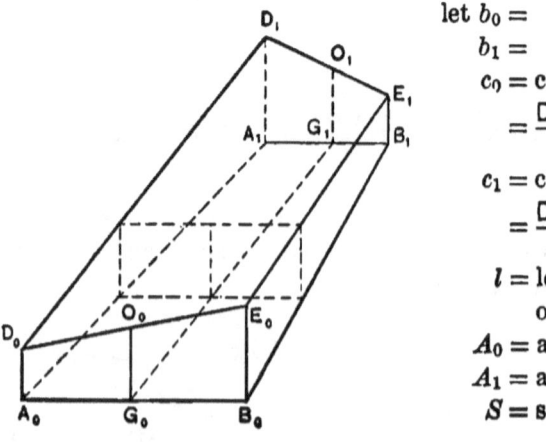

let $b_0 =$ base $= A_0B_0$
$b_1 =$ " $= A_1B_1$
$c_0 =$ center ht. $= O_0G_0$
$= \dfrac{D_0A_0 + E_0B_0}{2}$
$c_1 =$ center ht. $= O_1G_1$
$= \dfrac{D_1A_1 + E_1B_1}{2}$
$l =$ length (altitude) of section $= G_0G_1$
$A_0 =$ area of $D_0A_0B_0E_0$
$A_1 =$ area of $D_1A_1B_1E_1$
$S =$ solidity

Methods of Computing Earthwork. 131

Also use notation b_z, c_z, A_z for a section distant z from G_1.
Then

$A_0 = b_0 c_0 \qquad A_1 = b_1 c_1$

$b_z = b_1 + (b_0 - b_1)\dfrac{x}{l}$

$c_z = c_1 - (c_1 - c_0)\dfrac{x}{l} = c_1 + (c_0 - c_1)\dfrac{x}{l}$

$A_z = b_z c_z = \left[b_1 + (b_0 - b_1)\dfrac{x}{l}\right]\left[c_1 + (c_0 - c_1)\dfrac{x}{l}\right]$

$S = \displaystyle\int_0^l \left[b_1 + (b_0 - b_1)\dfrac{x}{l}\right]\left[c_1 + (c_0 - c_1)\dfrac{x}{l}\right]dx$

$= b_1 c_1 l + [b_1(c_0-c_1) + c_1(b_0-b_1)]\dfrac{l^2}{2\,l} + \dfrac{(b_0-b_1)(c_0-c_1)l^3}{3\,l^2}$

$= \dfrac{l}{6}\left\{\begin{array}{l} 6\,b_1c_1 + 3\,b_1c_0 + 3\,b_0c_1 + 2\,b_0c_0 \\ -\,3\,b_1c_1 - 2\,b_1c_0 - 2\,b_0c_1 \\ -\,3\,b_1c_1 \\ +\,2\,b_1c_1 \end{array}\right\}$

$S = \dfrac{l}{6}(2\,b_1c_1 + 2\,b_0c_0 + b_1c_0 + b_0c_1) \qquad (164)$

236. Apply the "Prismoidal Formula" to the same section. The base and center height of the middle section are: —

$b_m = \dfrac{b_0 + b_1}{2} \qquad\qquad c_m = \dfrac{c_0 + c_1}{2}$

$A_0 = b_0 c_0 \qquad\qquad A_1 = b_1 c_1$

$M = \dfrac{b_0 + b_1}{2} \times \dfrac{c_0 + c_1}{2} = $ area of middle section

$S = \dfrac{l}{6}(A_0 + 4M + A_1)$

$= \dfrac{l}{6}\,(b_0c_0 + b_0c_0 + b_0c_1 + b_1c_0 + b_1c_1 + b_1c_1)$

$= \dfrac{l}{6}(2\,b_1c_1 + 2\,b_0c_0 + b_1c_0 + b_0c_1) \qquad (165)$

This is the same as formula (164) found above to be correct for the warped surface. Therefore the "Prismoidal Formula" (163) applies to the section shown in § 235.

237. The sections of earthwork commonly used in railroad work are bounded not by perpendicular sides, but by inclined planes.

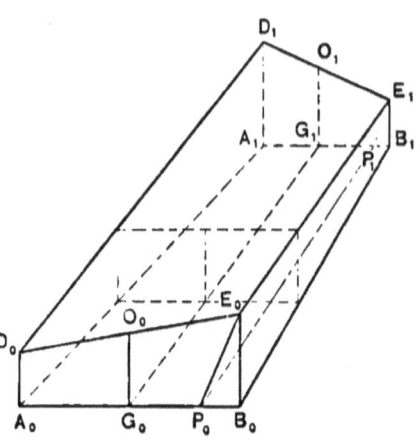

In the figure, suppose a plane to be passed through the line E_0E_1, cutting A_0B_0 at P_0 and A_1B_1 at P_1. The prismoidal formula applies to the solid $E_0P_0B_0B_1E_1P_1$ cut out by this plane, since this solid is a true prismoid. If the prismoidal formula applies to the entire solid, and also to the part cut out, it must apply to the remaining solid $D_0A_0P_0E_0E_1P_1A_1D_1$, and this represents in form one side of a regular three-level section of earthwork in which D_0A_0 represents the center height and E_0P_0 the slope.

If the prismoidal formula applies to the section upon one side of the center, it applies also to the other side, and so to the entire section.

238. The "*Prismoidal Formula*" is of wide application. Since it applies to prisms, wedges, pyramids, and to solids bounded by warped surfaces generated as described, it follows that it applies to any solid bounded by two parallel plane faces and defined by the surfaces generated by a right line moving upon the perimeters of these faces as directrices. It may also be stated here without demonstration that it also applies to the *frusta* of all solids generated by the revolution of a conic section as well as to the complete solids, for instance, the sphere.

The prismoidal formula is generally accepted as correct for the computation of earthwork and similar solids, and the measurements of a section of earthwork are taken so as to represent properly the surface of the ground if this be a warped surface of the sort described. The failure to use the prismoidal formula is explained often by the additional labor necessary for its use.

Methods of Computing Earthwork.

239. For "three-level" sections of earthwork, a result correct by the prismoidal formula may be secured, and the work simplified, by calculating the quantities first by the inexact method of "end areas," and then applying a *correction* which we may call "**The Prismoidal Correction.**"

Let S_e = solidity by end areas

$S_p =$ " " prismoidal formula

Then $C = S_e - S_p$ = prismoidal correction

In the figure, § 235,

$S_p =$ by formula (164) $= \dfrac{l}{6}(2 b_1 c_1 + 2 b_0 c_0 + b_1 c_0 + b_0 c_1)$

$S_e = \dfrac{l}{2}(b_1 c_1 + b_0 c_0) \quad = \dfrac{l}{6}(3 b_1 c_1 + 3 b_0 c_0)$

$$C = S_e - S_p = \dfrac{l}{6}(b_1 c_1 + b_0 c_0 - b_1 c_0 - b_0 c_1)$$

$$= \dfrac{l}{6}(b_1 - b_0)(c_1 - c_0)$$

Let $\quad D_0 A_0 = h_0' \qquad\qquad D_1 A_1 = h_1'$

$\qquad E_0' B_0 = h_0 \qquad\qquad E_1 B_1 = h_1$

Then $\quad C = \dfrac{l}{6}(b_1 - b_0)\left(\dfrac{h_1 + h_1'}{2} - \dfrac{h_0 + h_0'}{2}\right)$

$\qquad\quad = \dfrac{l}{12}(b_1 - b_0)(h_1 + h_1' - h_0 - h_0')$

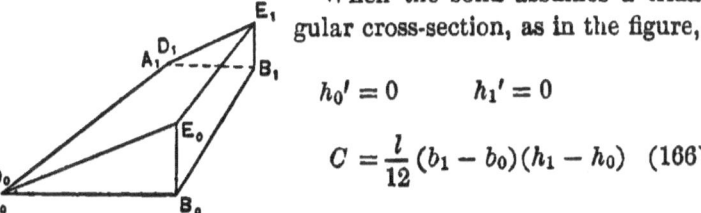

When the solid assumes a triangular cross-section, as in the figure,

$h_0' = 0 \qquad h_1' = 0$

$C = \dfrac{l}{12}(b_1 - b_0)(h_1 - h_0)$ \quad (166)

240. If any solid be divided into a number of solids each of triangular cross-section, the above correction may be applied to each such triangular solid, and the sum of the corrections will be the correction for the entire solid.

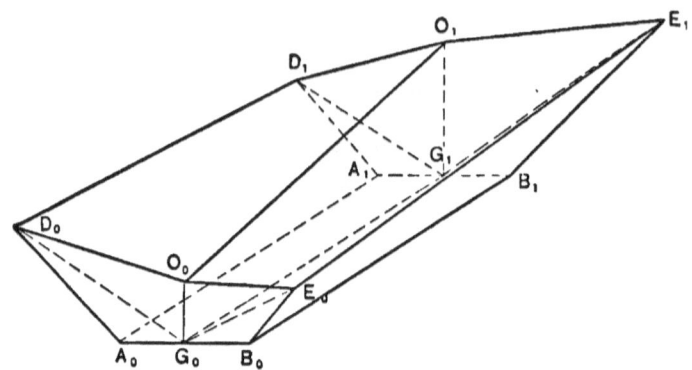

Let this figure represent a section of earthwork divided into three parts, as indicated by the lines D_0G_0, E_0G_0, D_1G_1, E_1G_1.

Then, for the solid $O_0D_0G_0E_0E_1G_1D_1O_1$,

$$C = \frac{l}{12}[(c_1 - c_0)(d_{l_1} - d_{l_0}) + (c_1 - c_0)(d_{r_1} - d_{r_0})]$$

$$= \frac{l}{12}(c_1 - c_0)(d_{l_1} + d_{r_1} - d_{l_0} - d_{r_0})$$

Let $D_1 = d_{l_1} + d_{r_1}$ and $D_0 = d_{l_0} + d_{r_0}$

$$C = \frac{l}{12}(c_1 - c_0)(D_1 - D_0)$$

For the solid $G_0B_0E_0E_1B_1G_1$,

(166) $\quad C = \frac{l}{12}\left(\frac{b_1}{2} - \frac{b_0}{2}\right)(h_{r_1} - h_{r_0})$

$$= \frac{l}{12}(0)(h_{r_1} - h_{r_0})$$

$$= 0$$

Similarly for the solid $A_0G_0D_0D_1G_1A_1$.

Hence for the entire solid $A_0B_0E_0O_0D_0D_1O_1E_1B_1A_1$.

$$C = \frac{l}{12}(c_1 - c_0)(D_1 - D_0) \qquad (167)$$

Methods of Computing Earthwork. 135

When $l = 100$

$$C = \frac{100}{12 \times 27}(c_1 - c_0)(D_1 - D_0)$$

$$= \frac{1}{3.24}(c_1 - c_0)(D_1 - D_0) \text{ in cu. yds.} \quad (168)$$

Since $C = S_e - S_p$

$$S_p = S_e - C \quad (169)$$

When $(c_1 - c_0)(D_1 - D_0)$ is *positive*, the correction C is to be *subtracted* from S_e.

When $(c_1 - c_0)(D_1 - D_0)$ is *negative*, the arithmetical value of C is to be *added* to S_e. The latter case seldom occurs in practice, except where C is very small, perhaps small enough to be neglected.

For a section of length l,

$$C_l = \frac{l}{100} C_{100}$$

$$S_{pl} = \frac{l}{100}(S_{e100} - C_{100})$$

241. In general, for sections of earthwork, the *prismoidal* correction as given above *applies only* when *the width of base is the same at both ends* of the section. There are certain special cases, however, which often occur, and which allow of the convenient use of this formula for prismoidal correction. Referring to the figure on p. 120, and the corresponding notes on p. 118, the correction can be correctly applied in the case of the *excavation* from Sta. $2 + 64$ to $2 + 76$ as follows:—

Compute S_e, and then apply C, using at

Sta. $2 + 64$ $D_0 = 23.2$
and at   Sta. $2 + 76$ $D_1 = 11.9 = d_{r_1}$

or the distance out on one side only. This may readily be demonstrated to be proper if the correction to the right of the center be taken, using formula (167), and the correction to the left using formula (166), and the two corrections (right and left) be added.

242. Formula (166) can also be used to find the correction for the triangular pyramids (for excavation Sta. 2+76 to 2+91, and embankment 2 + 64 to 2 + 76), *each end of the pyramid being considered to have a triangular section.* A much simpler way to find the correction for a pyramid is this,

$$C = S_e - S_p = \frac{1}{3} S_e$$

as may readily be shown to be true for any pyramid, since

$$S_e = A \frac{l}{2}$$

$$S_p = A \frac{l}{3}$$

$$C = S_e - S_p = A \frac{l}{6} = \frac{S_e}{3} \qquad (170)$$

243. In the case of regular "**Five-Level Sections**," as shown in the figure, p. 127, the prismoidal correction may be computed for each of the triangular masses bounded by

    1. AOB   2. OBE   3. OAD

In the case of AOB, the prismoidal correction will evidently be $= 0$, since $D_0 = 0 = D_1$, and therefore $D_0 - D_1 = 0$.

The correction for the mass bounded on one end by

$$\text{OBE} = C = \frac{l}{12}(f_{r_0} - f_{r_1})(d_{r_0} - d_{r_1})$$

and by

$$\text{OAD} = C = \frac{l}{12}(f_{l_0} - f_{l_1})(d_{l_0} - d_{l_1})$$

OBE and OAD differ but little from regular sections of earthwork in which $b = 0$.

244. In the case of "**Irregular Sections**," the prismoidal correction cannot with convenience be accurately employed. There are, however, several methods by which we may calculate a "prismoidal correction" which will be approximately correct.

For the purpose *only* of calculating the correction, either of the following methods may be employed:—

Methods of Computing Earthwork. 137

1. Neglect all intermediate heights, and figure correction from center and side heights.

2. Find level sections of equal area in each case, and figure correction, using the center heights and side distances of these level sections.

3. Having c and D of the irregular section, either

        (*a*) retain c and calculate D, or
        (*b*)   " D "   " c

for a "regular three-level" section of equal area, and use these values to calculate the correction.

4. Plat the cross-section on cross-section paper, and equalize by a line or lines drawn in the most advantageous direction, and from the c and D thus found compute the correction.

245. In relation to these methods : —

No. 1 is most rapid and least accurate.

" 2 is less exact than 3 in most cases, and probably no more rapid.

" 3 is recommended as nearly equal to 4 in accuracy, and far more rapid.

" 4 would yield the most accurate results.

The value of these approximate methods cannot be properly appreciated until certain rapid methods of computation are understood, as will be appreciated later.

The results obtained by the methods shown above are *approximate* only, but in most cases the resulting error would be small, or a small *fraction* only of the *entire correction*, which is itself generally small.

246. The method of calculating by averaging end areas and applying the *prismoidal correction* will be found much more rapid than to calculate the middle area and apply the *prismoidal formula* directly. There is another advantage of importance in favor of the use of the prismoidal correction ; in a majority of cases for sections used, the method of "end areas," is sufficiently accurate for all practical purposes, and from the use of the prismoidal correction the computer will soon learn to distinguish, by inspection merely, in what cases this correction need be applied.

247. III. Method of Middle Areas.

This consists in calculating the area of the middle section (*not the mean* of the end areas), and assuming the solidity to be that of a prism having a base equal to this middle area, and an altitude equal to the length of the section of earthwork.

Let M = middle area

l = length of section

Then $S = Ml$

This method is not exact. It gives results generally less than the correct solidity. It is not sufficiently rapid to recommend it.

248. IV. Method of Equivalent Level Sections.

This consists in finding level *end sections* of equal area with the actual end sections from these calculating the level *middle section*, assuming the top surface connecting the level end sections to be a plane; and then calculating the solidity of this prismoid by the *prismoidal formula*.

This method is not exact; it gives results less than the correct solidity. It is not sufficiently rapid to recommend it.

249. V. Method of Mean Proportionals.

This consists in assuming that the solid is the frustum of a pyramid, in which case all its sides would meet in one vertex. This method is not exact. It gives results always less than the correct solidity.

250. VI. Henck's Method.

In connection with the prismoidal formula, it was stated that the most common assumption was that the upper surface is a warped surface of a certain kind which was there described. *Henck's Method* assumes otherwise; that the upper surface is divided into plane surfaces by diagonals from the center height of one cross-section to the side height of the next, as shown in the figure, where the diagonals $O_1 E_0$ and $O_0 D_1$ are drawn.

Methods of Computing Earthwork. 139

Which way the diagonals shall be assumed to run is determined on the ground by the shape of the surface in each case.

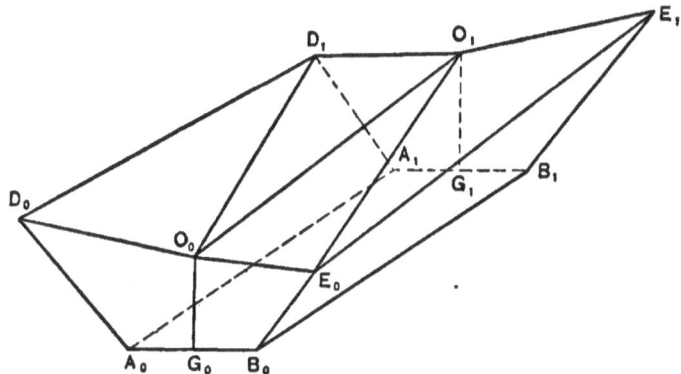

The diagonals O_1E_0 and O_0D_1 divide the surface into four plane surfaces, $D_0O_0D_1$, $O_0D_1O_1$, $O_0E_0O_1$, $E_0O_1E_1$.

251. Let this figure represent the right-hand side of a section of earthwork, with the diagonal assumed to run from E_0 to O_1.

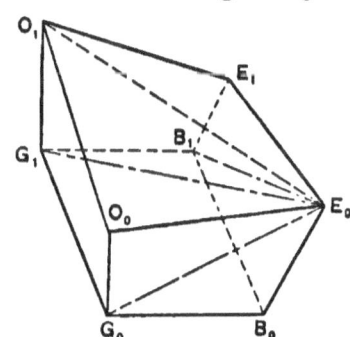

Join E_0 with G_0, B_1 and G_1.

The entire solid may then be considered as composed of three pyramids having their vertices in a common point E_0.

Using notation already familiar, the solidities of the three pyramids are as follows:—

$s_1 = $ area $G_0B_0B_1G_1 \times$ height at $E_0 \div 3$

$= \dfrac{1}{2}bl \times \dfrac{h_{r_0}}{3}$

$s_2 = $ area $G_0O_0O_1G_1 \times$ distance out to $E_0 \div 3$

$= \dfrac{c_0 + c_1}{2}l \times \dfrac{d_{r_0}}{3}$

$s_3 = $ area $O_1G_1B_1E_1 \times$ length of section $\div 3$

$= \dfrac{1}{2}\left(\dfrac{b}{2}h_{r_1} + c_1d_{r_1}\right) \times \dfrac{l}{3}$

140   *Railroad Curves and Earthwork.*

Let S_r = solidity of this right half of section

$S_r = s_1 + s_2 + s_3$

$= \frac{1}{6} blh_{r_0} + \frac{1}{6} d_{r_0} lc_0 + \frac{1}{6} d_{r_0} lc_1 + \frac{1}{6} \frac{bh_{r_1}l}{2} + \frac{1}{6} c_1 d_r l$

$= \frac{l}{6}\left[bh_{r_0} + \frac{bh_{r_1}}{2} + c_0 d_{r_0} + c_1 d_{r_1} + c_1 d_{r_0} \right].$

Let C = center height touched by diagonal

H = side   "   "   "   "

D = distance out to side height touched by diagonal

$S_r = \frac{l}{6}\left[\frac{b}{2}(h_{r_0} + h_{r_1} + H_r) + c_0 d_{r_0} + c_1 d_{r_1} + C_r D_r \right]$

Then $S_l = \frac{l}{6}\left[\frac{b}{2}(h_{l_0} + h_{l_1} + H_l) + c_0 d_{l_0} + c_1 d_{l_1} + C_l D_l \right]$

= solidity for left half of section

$S = S_r + S_l = \frac{l}{6}\left[\frac{b}{2}(h_{r_0} + h_{l_0} + h_{r_1} + h_{l_1} + H_r + H_l) + c_0(d_{r_0} + d_{l_0}) \right.$

$\left. + c_1(d_{r_1} + d_{l_1}) + C_r D_r + C_l D_l \right]$ (171)

252. An example will further show the application of this method.

Sta.	Surf. Elev.	Grade Elev.	Cross-Sections		
1	123.0	121.00	$\frac{13.0}{+3.0}$	$+2.0$	$\frac{11.0}{+1.0}$
0	123.0	120.00	$\frac{15.0}{+5.0}$	$+3.0$	$\frac{11.0}{+1.0}$

The notes show the direction of the diagonal as taken on the ground.

In this case $b = 20$ $s = 1$ to 1

Methods of Computing Earthwork. 141

253. Henck gives note-book and calculations in this form : —

STA.	d_l	h_l	c	h_r	d_r	d_l+d_r	$(d_l+d_r)c$	$D_r C_r$	$D_l C_l$
1	13.0	3.0	2.0	1.0	11.0	24.0	48.0		
0	15.0	5.0	3.0	1.0	11.0	26.0	78.0	22.0	39.0
		8.0	=		8.0		22.0		
					4.0		39.0		
					14.0 $\times \frac{20}{2}$ =		140.0		
							6)327.0		
							5450 (cu. ft.)		

254. The calculations could be conveniently made however from the notes as now generally taken, as is shown below : —

$$
\begin{array}{cccc}
1 & \dfrac{13.0}{+3.0} & +2.0 & \dfrac{11.0}{+1.0} \quad 24.0 \times 2.0 = 48.0 \\
0 & \dfrac{15.0}{+5.0} & +3.0 & \dfrac{11.0}{+1.0} \quad 26.0 \times 3.0 = 78.0 \\
& & & 2.0 \quad\quad 11.0 \times 2.0 = 22.0 \\
& 8.0 & & 8.0 \quad\quad 13.0 \times 3.0 = 39.0 \\
& & & 4.0 \\
& & & \overline{14.0} \times \dfrac{20}{2} \\
& & & = 140.0 \\
& & & \overline{6)327.0} \\
& & & 5450 \text{ (cu. ft.)}
\end{array}
$$

255. The work of computation would not, in either of these cases, properly be done in the field note-book, but rather in a calculation book, or other suitable place.

For a *series* of cross-sections, Henck systematizes the work, and reduces the labor noticeably from what is shown here. (See Henck's Field Book.) Henck's method is strictly accurate, upon the assumption made as to the upper surface. In general railroad practice, most engineers prefer to assume the upper surface a warped surface of the sort described.

Henck's method is less rapid than that of averaging end areas and applying the prismoidal correction.

142   Railroad Curves and Earthwork.

256. The method of averaging end areas and applying the prismoidal correction appears in point of accuracy and rapidity to meet the requirements of modern railroad practice.

Some engineers whose opinions are entitled to careful consideration object to the use of the prismoidal formula or prismoidal correction in any form, some as an unnecessary refinement, and some on the ground that certain practical considerations render the results nearer the truth when the method of averaging end areas is used without applying the prismoidal correction. Probably the greater part of the best engineering practice favors the use of the prismoidal correction.

257. Example.

Showing a comparison of various methods of calculating earthwork.

Notes of excavation.   Base 24.   Slope 1½ to 1.

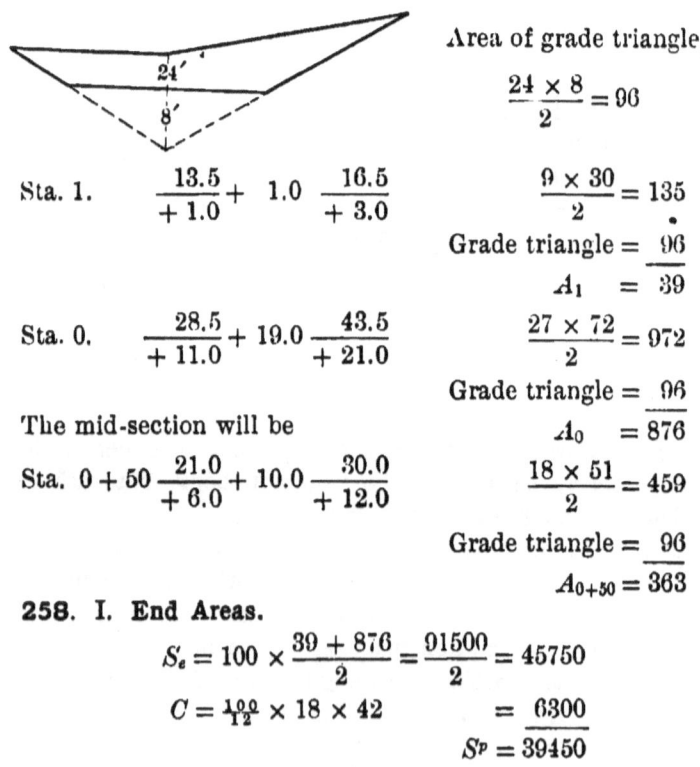

Area of grade triangle,

$$\frac{24 \times 8}{2} = 96$$

Sta. 1. $\frac{13.5}{+1.0} + 1.0 \frac{16.5}{+3.0}$

$$\frac{9 \times 30}{2} = 135$$
Grade triangle = 96
A_1 = 39

Sta. 0. $\frac{28.5}{+11.0} + 19.0 \frac{43.5}{+21.0}$

$$\frac{27 \times 72}{2} = 972$$
Grade triangle = 96
A_0 = 876

The mid-section will be

Sta. $0 + 50$ $\frac{21.0}{+6.0} + 10.0 \frac{30.0}{+12.0}$

$$\frac{18 \times 51}{2} = 459$$
Grade triangle = 96
$A_{0+50} = 363$

258. I. End Areas.

$$S_e = 100 \times \frac{39 + 876}{2} = \frac{91500}{2} = 45750$$

$$C = \tfrac{100}{12} \times 18 \times 42 \qquad = 6300$$

$$S^p = 39450$$

Error of $S_e = + 6300 = 16$ per cent

Methods of Computing Earthwork. 143

259. II. Prismoidal Formula.

$$39 = A_1$$
$$363 \times 4 = 1452 = 4 A_{0+50}$$
$$876 = A_0$$
$$6\overline{)236700}$$
$$39450 = S_p$$

260. III. Middle Areas.

$$S_m = 363 \times 100 = 36300$$
$$S_m \text{ error} = -3150 = 8 \text{ per cent}$$
$$S_e \quad \text{''} \quad = +6300$$

261. IV. Equivalent Level Sections.

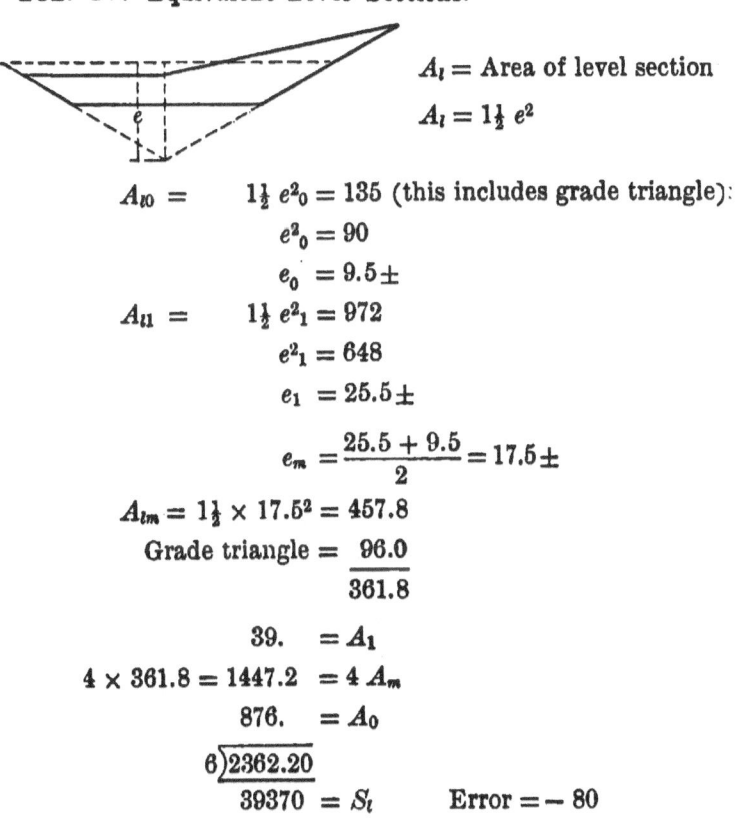

A_l = Area of level section
$A_l = 1\tfrac{1}{2} e^2$

$A_{l0} = \quad 1\tfrac{1}{2} e^2{}_0 = 135$ (this includes grade triangle):
$\qquad e^2{}_0 = 90$
$\qquad e_0 = 9.5\pm$
$A_{l1} = \quad 1\tfrac{1}{2} e^2{}_1 = 972$
$\qquad e^2{}_1 = 648$
$\qquad e_1 = 25.5\pm$

$$e_m = \frac{25.5 + 9.5}{2} = 17.5\pm$$

$A_{lm} = 1\tfrac{1}{2} \times 17.5^2 = 457.8$
Grade triangle = $\underline{96.0}$
$\qquad\qquad\qquad 361.8$

$$39. = A_1$$
$$4 \times 361.8 = 1447.2 = 4 A_m$$
$$876. = A_0$$
$$6\overline{)2362.20}$$
$$39370 = S_l \qquad \text{Error} = -80$$
$$\qquad\qquad\qquad = 0.2 \text{ per cent.}$$

262. V. Mean Proportionals.

$A_0 = 39$ $39. = A_0$
$A_1 = 876$ $184.8 = \sqrt{A_0 A_1}$
$A_0 A_1 = 34164$ $876. = A_1$
$\sqrt{A_0 A_1} = 184.8$ $3\overline{)1099.80}$
               $36660. = S_{\sqrt{\ }} = \dfrac{l}{3}(A_0 + \sqrt{A_0 A_1} + A_1)$

Error − 2790. = 7 per cent

263. VI. Henck's Method.

ONE SYSTEM OF DIAGONALS AS SHOWN IN NOTES.

STA.	d_l	h_l	c	h_r	d_r	d_l+d_r	$(d_l+d_r)c$	$D'C'$	DC
1	13.5	1.0	/1.0\	3.0	16.5	30.0	30.0		
0	28.5	11.0/	19.0	\21.0	43.5	72.0	1368.0	7	2.
		12.0		24.0			72.0		
				12.0			816.0		
				11.0			$6\overline{)2286.}$		
				21.0			$38100 = S_h$		
				68.0					

$\dfrac{b}{2} = \dfrac{12.}{816.}$ $S_p - S_h = +1350$
              $= 3$ per cent

OPPOSITE SYSTEM OF DIAGONALS.

STA.	d_l	h_l	c	h_r	d_r	d_l+d_r	$(d_l+d_r)c$	DC'	DC
1	13.5	1.0\	1.0	/3.0	16.5	30.0	30.0		
0	28.5	11.0	\19.0/	21.0	43.5	72.0	1368.0	57	0.
				24.0			570.0		
				12.0			480.0		
				1.0			$6\overline{)2448.0}$		
				3.0			$40800 = S_{h'}$		
				40.0			$38100 = S_h$		

$\dfrac{b}{2} = \dfrac{12.}{480.}$ $2\overline{)78900}$
                     39450

$S_p - S_{h'} = -1350 = 3$ per cent   Mean value $= S_p$

CHAPTER XIII.

SPECIAL PROBLEMS.

264. Correction for Curvature.

In the case of a curve, the ends of a section of earthwork are not parallel, but are in each case normal to the curve. In calculating the solidity of a section of earthwork, we have heretofore assumed the ends parallel, and for curves this is equivalent to taking them perpendicular to the chord of the curve between the two stations.

Then, as shown in Fig. 1 (where IG and GT are center-line chords), the solidity (as above) of the sections IG and GT will be too great by the wedge-shaped mass RGP, and too small by

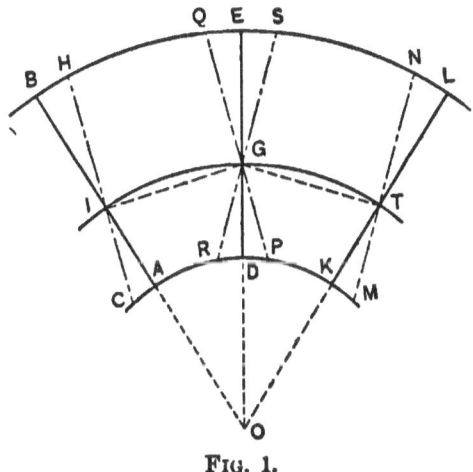

Fig. 1.

QGS. When the cross-sections on each side of the center are equal, these masses balance each other. When the cross-section on one side differs much in area from that on the other, the correction necessary may be considerable.

In Fig. 2, use c, h_l, h_r, d_l, d_r, b, s, as before.

Let D = degree of curve. Make BL = AD, and join OL.

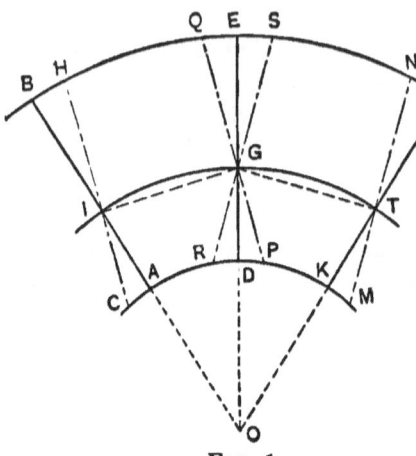

FIG. 1.

Then ODAG balances OLBG, and there remains an unbalanced area OLE.

Draw OKP parallel to AB.

By the "Theorem of Pappus" (see Lanza, Applied Mechanics), "If a plane area lying wholly on the same side of a straight line in its own plane revolves about that line, and thereby generates a solid of revolution, the volume of the solid thus generated is equal to the product of the revolving area and of the path described by the center of gravity of the plane area during the revolution."

The correction for curvature, or the solidity, developed by this triangle OLE (Fig. 2) revolving about OG as an axis will be its area × the distance described by its center of gravity. The distance out (horizontal) to the center of gravity from the axis (center line) will be two thirds of the mean of the distances out to E and to L, or

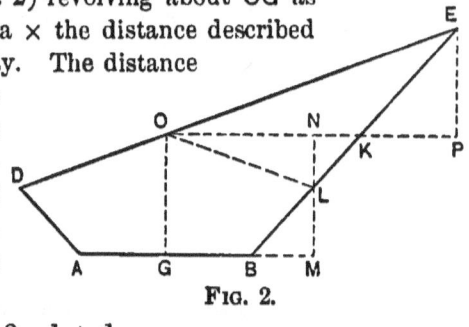

FIG. 2.

$$= \frac{2}{3} \cdot \frac{d_l + d_r}{2}$$

and the distance described will be

$$\frac{2}{3} \cdot \frac{d_l + d_r}{2} \times \text{angle QGS}$$

The area   OLE = OK × $\dfrac{\text{NL} + \text{PE}}{2}$

$$= \left(\frac{b}{2} + sc\right)\frac{h_r - h_l}{2}$$

Special Problems. 147

Therefore the correction for curvature,

$$C = \left(\frac{b}{2} + sc\right) \cdot \frac{h_r - h_l}{2} \cdot \frac{d_r + d_l}{3} \times \text{angle QGS}$$

When IG, GT are each a full station, or 100 ft. in length,

$$\text{QGS} = D$$

$$C = \left(\frac{b}{2} + sc\right) \cdot \frac{h_r - h_l}{2} \cdot \frac{d_r + d_l}{3} \times \text{angle } D$$

arc $1° = .01745$

$$C = \left(\frac{b}{2} + sc\right) \frac{h_r - h_l}{2} \times \frac{d_r + d_l}{3} \times 0.01745 \, D$$

$$= \left(\frac{b}{2} + sc\right)(h_r - h_l)(d_r + d_l) \times 0.00291 \, D \text{ (cu. ft.)} \quad (172)$$

$$= \left(\frac{b}{2} + sc\right)(h_r - h_l)(d_r + d_l) \times 0.00011 \, D \text{ (cu. yds.)} \quad (173)$$

265. When IG or GT, or both, are less than 100 ft., let

$$\text{IG} = l_0 \quad \text{and} \quad \text{GT} = l_1$$

Then $\quad \text{QGE} = \dfrac{l_0}{100} \times \dfrac{D}{2}$ and $\text{SGE} = \dfrac{l_1}{100} \times \dfrac{D}{2}$

$$\text{QGS} = \frac{l_0 + l_1}{200} D$$

$$C = \left(\frac{b}{2} + sc\right)(h_r - h_l)(d_r + d_l) \frac{l_0 + l_1}{200} \times 0.00011 \, D \text{ (cu. yds.)} \quad (174)$$

266. The correction C is to be added when the greater area is on the outside of the curve, and subtracted when the greater area is on the inside of the curve. When the center height is 0, as in Fig. 3, we may consider this a regular section in which $c = 0$, $h_l = 0$, and $d_l = \dfrac{b}{2}$; then

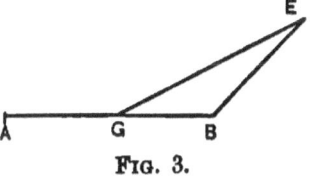

Fig. 3.

$$C = \frac{b}{2} \times h_r \times \left(d_r + \frac{b}{2}\right) \frac{l_0 + l_1}{200} \times 0.00011 \, D \text{ (cu. yds.)} \quad (175)$$

In the case of an irregular section, as shown in Fig. 4, the area and distance to center of gravity (for example, of OHEML) may be found by any method available, and the correction

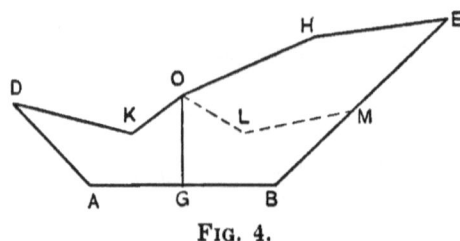

Fig. 4.

figured accordingly. The correction for curvature is, in present railroad practice, more frequently neglected than used. Nevertheless, its amount is sufficient in many cases to fully warrant its use.

267. Opening in Embankment.

Where an opening is left in an embankment, there remains outside the regular sections the mass DEKHF.

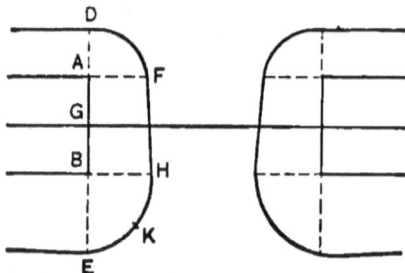

This must be calculated in 3 pieces, ADF, BEKH, ABHF.

Let b = base = AB
 d_r = distance out right
 d_l = distance out left
 $\left.\begin{array}{l}p_r = BH \\ p_l = AF\end{array}\right\}$ taken parallel to center line
 $\left.\begin{array}{l}f_r = \\ f_l =\end{array}\right\}$ heights at $\left\{\begin{array}{l}B \\ A\end{array}\right.$
 s_1 = solidity ADF
 s_2 = BEKH
 s_3 = ABHF

Special Problems. 149

Then (approximately) following the "Theorem of Pappus,"
s_1 = mean of triangles AD and AF × distance described by center of gravity.

$$\text{mean area} = \frac{\text{area AD} + \text{area AF}}{2} = \frac{\frac{f_l}{2}\left(d_l - \frac{b}{2}\right) + \frac{f_l}{2} p_l}{2}$$

$$= \frac{f_l}{2} \times \frac{1}{2}\left(d_l + p_l - \frac{b}{2}\right)$$

The distance described by center of gravity is found thus:—

$$\text{length AD} = d_l - \frac{b}{2}, \text{ and AF} = p_l$$

$$\text{mean length} = \frac{\text{length AD} + \text{length AF}}{2}$$

$$= \frac{1}{2}\left(d_l + p_l - \frac{b}{2}\right)$$

$$\text{distance out to center of gravity} = \frac{1}{3} \cdot \frac{1}{2}\left(d_l + p_l - \frac{b}{2}\right)$$

$$\text{distance described by center of gravity} = \frac{1}{6}\left(d_l + p_l - \frac{b}{2}\right)\frac{\pi}{2}$$

$$s_1 = \frac{f_l}{2} \times \frac{1}{2}\left(d_l + p_l - \frac{b}{2}\right) \times \frac{1}{6}\left(d_l + p_l - \frac{b}{2}\right)\frac{\pi}{2}$$

$$= f_l\left(d_l + p_l - \frac{b}{2}\right)^2 \frac{\pi}{48}$$

$$= \frac{f_l}{15}\left(d_l + p_l - \frac{b}{2}\right)^2 \text{ nearly} \tag{176}$$

$$s_2 = \frac{f_r}{15}\left(d_r + p_r - \frac{b}{2}\right)^2 \text{ nearly} \tag{177}$$

$$s_3 = \frac{\text{area AF} + \text{area BH}}{2} \times \text{AB}$$

$$= \frac{(f_l p_l + f_r p_r)b}{4} \tag{178}$$

268. The work of deriving formulas (176) and (177) is approximate throughout, but the total quantities involved are in general not large, and the error resulting would be unimportant.

There seems to be no method of accurately computing this solidity which is adapted to general railroad practice.

269. Borrow-Pits.

In addition to the ordinary work of excavation and embankment for railroads, earth is often "borrowed" from outside the limits of the work proper; and in such excavations called "borrow-pits," it is common to prepare the work by dividing the surface into squares, rectangles, or triangles, taking levels at every corner upon the original surface; again, after the excavation of the borrow-pit is completed, the points are reproduced and levels taken a second time. The excavation is thus divided into a series of vertical prisms having square, rectangular, or triangular cross-sections. These prisms are commonly truncated top and bottom. The lengths or altitudes of the vertical edges of these prisms are given by the difference in levels taken,

1st, on the original surface, and

2d, after the excavation is completed.

This method of measurement is very generally used, and for many purposes.

270. Truncated Triangular Prisms.

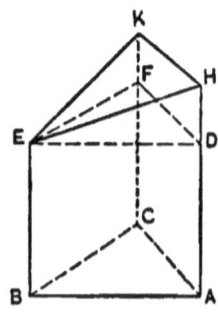

Let A = area of right section EFD of a truncated prism, the base ABC being a right section

h_1 = height AH

h_2 = " BE

h_3 = " CK

a = altitude of triangle EFD dropped from E to FD

Let S = solidity of prism ABCKHE

s_l = " " " ABCFDE

s_u = " " pyramid FDEHK

Special Problems. 151

Then $\quad s_l = A \times AD = A \times \dfrac{3\,AD}{3} = A \times \dfrac{AD + BE + CF}{3}$

$s_u = \text{area DFKH} \times \dfrac{a}{3}$

$\quad = \dfrac{KF + HD}{2} \times FD \times \dfrac{a}{3}$

$\quad = \dfrac{KF + HD}{3} \times FD \times \dfrac{a}{2}$

$\quad = \dfrac{KF + HD}{3} \times A$

$S = s_l + s_u = A\left(\dfrac{AD + BE + CF}{3} + \dfrac{KF + HD}{3}\right)$

$\quad = A\dfrac{(AD + HD) + BE + (CF + KF)}{3}$

$\quad = A\dfrac{h_1 + h_2 + h_3}{3} \qquad (179)$

If the prism be truncated top and bottom, the same reasoning holds and the same formula applies.

271. Truncated Rectangular Prism.

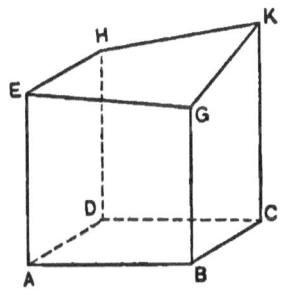

Let A = area of right section ABCD of a rectangular prism truncated on top (base is ABCD)

h_1 = height AE

$h_2 = \quad\text{``}\quad$ BG

$h_3 = \quad\text{``}\quad$ KC

$h_4 = \quad\text{``}\quad$ HD

S = solidity of prism

$b = AD = BC$

$a = AB = DC$

Then using method of end areas,

$$S = \frac{AEHD + BGKC}{2} \times a$$

$$= \frac{b\frac{h_1 + h_4}{2} + b\frac{h_2 + h_3}{2}}{2} \times a$$

$$= ab\,\frac{h_1 + h_2 + h_3 + h_4}{4}$$

$$S = A\,\frac{h_1 + h_2 + h_3 + h_4}{4} \text{ (cu. ft.)} \qquad (180)$$

$$S = \frac{A}{27} \cdot \frac{h_1 + h_2 + h_3 + h_4}{4} \text{ (cu. yds.)} \qquad (181)$$

We may find S, correct by the prismoidal formula, if we apply the prismoidal correction. The prismoidal correction $C' = 0$, since $D_0 - D_1 = 0$ (or in this case AD − BC = 0). The formula therefore remains unchanged. It is evident from this, then, that the solution holds good, and the formula is correct, not only when the surface EHKG is a plane, but also when it is a warped surface generated by a right line moving always parallel to the plane ADHE, and upon EG and HK as directrices.

Some engineers prefer to cross-section in rectangles of base $15' \times 18'$. In this case

$$S = \frac{15' \times 18'}{27} \cdot \frac{h_1 + h_2 + h_3 + h_4}{4} \text{ (cu. yds.)}$$

$$= 10\,\frac{h_1 + h_2 + h_3 + h_4}{4} \text{ (cu. yds.)} \qquad (182)$$

Other convenient dimensions will suggest themselves, as

$10' \times 13.5'$ or $20' \times 13.5'$ or $20' \times 27'$

By this method the computations are rendered slightly more convenient; but the size of the cross-section, and the shape, whether square or rectangular, should depend on the topography. The first essential is accuracy in results, the second is simplicity and economy in field-work, and ease of computation should be subordinate to both of these considerations.

272. Assembled Prisms.

In the case of an assembly of prisms of equal base, it is not necessary to separately calculate each prism, but the solidity of a number of prisms may be calculated in one operation.

In the prism B,
$$S_B = A\frac{a_2 + a_3 + b_3 + b_2}{4}$$

$$S_C = A\frac{a_3 + a_4 + b_4 + b_3}{4}, \text{ etc.}$$

From inspection it will be seen, taking A as the common area of base of a single prism, and taking the sum of the solidities, that the heights a_2, a_5 enter into the calculation of

one prism only; a_3, a_4 into two prisms each; b_1, b_6 one only; b_2, b_5 into three prisms; b_3, b_4 into four prisms; and similarly throughout.

Let $t_1 =$ sum of heights common to one prism
$t_2 = $ " " " " " two prisms
$t_3 = $ " " " " " three "
$t_4 = $ " " " " " four "

Then the total solidity,

CHAPTER XIV.

EARTHWORK TABLES.

273. The calculation of quantities can be much facilitated by the use of suitably arranged "Earthwork Tables."

For regular "Three-Level Sections" very convenient tables can be calculated upon the following principles or formulas:—

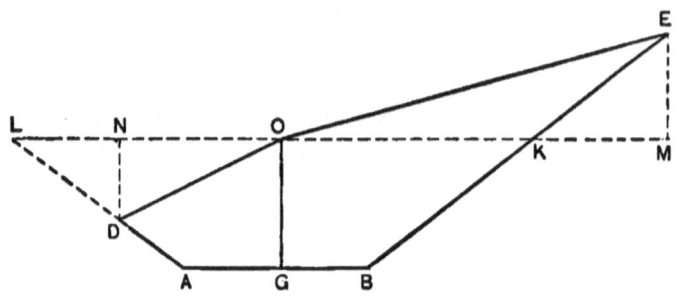

Use notation as before for

$$c, h_l, h_r, d_l, d_r, s, l, A, S$$

Then $\quad A = $ ABKL $+$ OKE $-$ ODL

$$= c(b + sc) + \frac{\text{OK} \times \text{EM}}{2} - \frac{\text{OL} \times \text{ND}}{2}$$

$$= c(b + sc) + \frac{\text{OK}}{2}(\text{EM} - \text{ND})$$

$$= c(b + sc) + \frac{1}{2}\left(\frac{b}{2} + sc\right)(h_l - c - c + h_r)$$

$$A = c(b + sc) + \frac{1}{2}\left(\frac{b}{2} + sc\right)(h_l + h_r - 2c) \qquad (185)$$

For a prism of base A and $l = 50$,

$$S = 50 A \text{ (cu. ft.)} = \frac{50}{27} A \text{ (cu. yds.)}$$

$$S = \frac{50}{27} c(b + sc) + \frac{25}{27}\left(\frac{b}{2} + sc\right)(h_l + h_r - 2c) \text{ (cu. yds.)} \qquad (186)$$

274. For cross-sections of a given base and slope, that is, given b and s constant, we may calculate for successive values of c, and tabulate values of L and K as follows: —

c	L	K
	$\dfrac{50}{27} c(b + sc)$	$\dfrac{25}{27}\left(\dfrac{b}{2} + sc\right)$

L represents the solidity for the *level section*.

K is for use as a correction. The formula then adapts itself to this table for any desired values of c, h_l, h_r.

$$S = L + K(h_l + h_r - 2c) \qquad (164)$$

Having found for successive stations S_0 and S_1 (each for a prism $l = 50$), then for the *full section* by "end areas,"

$$S_{100} = S_0 + S_1$$

for
$$S_{100} = \frac{A_0 + A_1}{2} \cdot \frac{100}{27} = \frac{50 A_0}{27} + \frac{50 A_1}{27}$$

$$S_{100} = \qquad\qquad S_0 + S_1 \qquad (187)$$

275. When l is less than 100,

$$S_l = (S_0 + S_1)\frac{l}{100} \qquad (188)$$

For level sections $\qquad h_l = h_r = c$

$$h_l + h_r - 2c = 0$$

and the formula

$$S = L + K(h_l + h_r - 2c)$$

becomes $\qquad S = L \qquad (189)$

for level sections, and the quantities for any given values of c can be directly taken from column L without any correction from column K.

In preliminary estimates, or wherever center heights only are used, such tables are rapidly used.

276. Tables may be found at the back of the book, pages 190, 191, calculated for

1. $b = 20$ $s = 1\frac{1}{2}$ to 1
2. $b = 14$ $s = 1\frac{1}{2}$ to 1

An example will illustrate their use,

$b = 14$ $s = 1\frac{1}{2}$ to 1

Notes:—

Sta. 1 $\dfrac{13.0}{-4.0}$ -3.7 $\dfrac{12.4}{-3.6}$

0    $\dfrac{10.6}{-2.4}$ -2.5 $\dfrac{10.3}{-2.2}$

Calculations:—

3.7 $L = 134.0$ $K = 11.6$ $h_l + h_r = 7.6$
       $+\ \ \ 2.3$ $\ \ 0.2$ $2c = 7.4$
       $S_1 = 136.3$ $\overline{2.32}$ $+ 0.2$

2.5 $L =\ \ 82.2$ $K = 10.0$ $h_l + h_r = 4.6$
      $-\ \ \ \ 4.0$ $\ \ 0.4$ $2c = 5.0$
      $S_0 =\ \ 78.2$ $\overline{4.00}$ $-\ 0.4$

$S_{100} = S_1 + S_0 = 214.5$

277. There is also at the back of this book a "Table of Prismoidal Correction" calculated by the formula

$$C = \frac{1}{3.24}(c_0 - c_1)(D_0 - D_1)$$

In the example above

$c_0 - c_1 = 2.5 - 3.7 = -1.2$

$D_0 - D_1 = 20.9 - 25.4 = -4.5$

From Table find opp. 4.5 for 1   1.39
                                    2   2.78(10
                                    2   0.28 $-$ 0.28
                                 10   $C = \overline{1.67}$

$S_{100} = S_e = 214.5$
$C =\ \ \ \ \underline{1.7}$
$S_p =\ 212.8$

Earthwork Tables.

278. When the section is less than 100 ft. in length, the prismoidal correction is made before multiplying by $\frac{l}{100}$; that is,

$$S_l = (S_0 + S_1 - C)\frac{l}{100} \tag{190}$$

279. Tables based upon these formulas have been published as follows: —

"The Civil Engineer's Excavation and Embankment Tables," by Clarence Pullen and Charles C. Chandler, published by the "J. M. W. Jones Stationery and Printing Co.," Chicago.

Tables are calculated for $b = 12$, 14, 16, 18, 20

$$s = \tfrac{1}{4}, \quad \tfrac{1}{2}, \quad 1, \quad 1\tfrac{1}{2}$$

Tables of a similar kind, but calculated so that

$$S_{100} = \frac{S_1 + S_0}{2}$$

are Hudson's Tables, published by John Wiley & Sons, New York.

280. For general calculation adapted both to regular "Three-Level Sections" and to "Irregular Sections," tables can be calculated upon the following principles and formulas: —

These tables are, in effect, tables of "Triangular Prisms," in which, having given (in feet) the base B and altitude a of any triangle, the tables give the solidity (in cu. yds.) for a prism of length $l = 50$; that is,

$$S = \frac{aB}{2} \cdot \frac{50}{27} = \frac{50}{54} aB \tag{191}$$

Whenever the calculation can be brought into the form $S = \frac{50}{54} aB$, the result can be taken directly from the table.

281. Tables of this kind are "Allen's Tables for Earthwork Computation," by the author of this book, and for sale by D. Van Nostrand Co., 23 Murray St., New York. A sample page is shown at the end of this book, page 194. Convenient tables of the same kind, but arranged differently for use, are "Tables for the Computation of Railway and other Earthwork," by C. L. Crandall, C.E., Ithaca, New York.

282. In both tables the formula $S = \dfrac{50}{54} aB$ takes form thus, $S = \dfrac{50}{54} \times \text{width} \times \text{height}$, and the tables are arranged as below.

	HEIGHTS.
WIDTHS	$\dfrac{50}{54} \text{ width} \times \text{height}$

The application to "Three-Level Sections" is as follows:—
We have formula (162), p. 126,

$$A = \left(c + \frac{b}{2s}\right)\frac{D}{2} - \frac{b^2}{4s}$$

and for a prism 50 ft. in length ($l = 50$)

$$S = \frac{50}{27} A = \frac{50}{54}\left(c + \frac{b}{2s}\right)D - \frac{50}{54} \cdot \frac{b}{2s} \cdot b$$

or S is the sum of two quantities, each of which is in proper form for the use of the tables.

For cross-sections of a given base and slope (b and s constants), $\dfrac{b}{2s}$ is a constant, and also $\dfrac{50}{54} \cdot \dfrac{b}{2s} \cdot b$ is constant.

We may then calculate once for all $\dfrac{b}{2s}$, and call this B (a constant).

Also $\dfrac{50}{54} \cdot \dfrac{b}{2s} \cdot b$, and call this a constant E.

Then
$$S = \frac{50}{54}(c + B)D - E \qquad (192)$$

In using the tables, $c + B = \text{height}$

$$D = \text{width}$$

As in the previous tables, having found S_0 and S_1,

$$S_{100} = S_0 + S_1$$

and
$$S_l = (S_0 + S_1)\frac{l}{100}$$

Earthwork Tables. 159

283. Example. Allen's Tables for Earthwork Computation.

Notes: —

Sta. 1 $\dfrac{9.1}{-2.4}$ -1.2 $\dfrac{7.3}{-1.2}$

Sta. 0 $\dfrac{8.8}{-2.2}$ -0.7 $\dfrac{6.4}{-0.6}$

$b = 11$ $s = 1\tfrac{1}{2}$ to 1

$$\dfrac{b}{2s} = 3.7 = B$$

Grade triangle, $\dfrac{50}{54} \times 3.7 \times 11$

Under height 3.7, find
    1 = 3.43   10. = 34.3
    1 = 3.43   1. = $\underline{3.4}$
                    $E = 37.7$

Station 1. $c = 1.2$
              $B = \underline{3.7}$
        height = 4.9

$D = 9.1 + 7.3 = 16.4$

Under height 4.9, find
    1 = 4.54   10. = 45.4
    6 = 27.22   6. = 27.2
    4 = 18.15   .4 = $\underline{1.8}$
                    74.4
                    $E = \underline{37.7}$
                    $S_1 = 36.7$

Station 0. $c = 0.7$
              $B = \underline{3.7}$
        height = 4.4

$D = 8.8 + 6.4 = 15.2$

Under height 4.4, find
    1 = 4.07   10. = 40.7
    5 = 20.37   5. = 20.4
    2 = 8.15   .2 = $\underline{0.8}$
                    61.9
                    $E = \underline{37.7}$
                    $S_0 = 24.2$

$$S = S_1 + S_0 = 60.9$$

284. Irregular Sections.

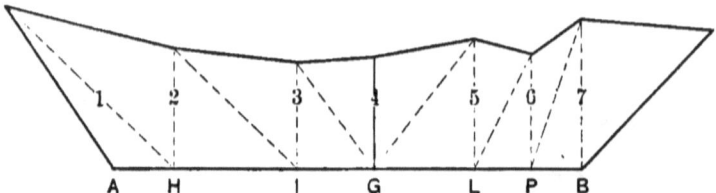

An "Irregular Section" can be divided into triangular parts, as in the figure. Taking generally two triangular parts together for purposes of calculation, we have

$$A_1 = \frac{h_l \times (AG - d_H)}{2} \qquad s_1 = \frac{50}{54} h_l(AG - d_H)$$

$$A_2 = \frac{h_H \times (d_l - d_I)}{2} \qquad s_2 = \frac{50}{54} h_H(d_l - d_I)$$

$$A_3 = \frac{h_I \times (d_H - 0)}{2} \qquad s_3 = \frac{50}{54} h_I d_H$$

$$A_4 = \frac{c \times (d_I + d_L)}{2} \qquad s_4 = \frac{50}{54} c(d_I + d_L)$$

$$A_5 = \frac{h_L \times (d_P - 0)}{2} \qquad s_5 = \frac{50}{54} h_L d_P$$

$$A_6 = \frac{h_P \times (d_B - d_L)}{2} \qquad s_6 = \frac{50}{54} h_P(d_B - d_L)$$

$$A_7 = \frac{h_B \times (d_r - d_P)}{2} \qquad s_7 = \frac{50}{54} h_B(d_r - d_P)$$

$$S = s_1 + s_2 + s_3 + s_4 + s_5 + s_6 + s_7 \qquad (193)$$

$$S_{100} = S_0 + S_1$$

$$S_l = (S_0 + S_1)\frac{l}{100}$$

285. The calculation of Irregular Sections in rough country becomes very laborious unless the best methods are used, and this process should be thoroughly understood.

CHAPTER XV.

EARTHWORK DIAGRAMS.

286. Computations of earthwork may also be made by means of diagrams from which results may be read by inspection merely.

The principle of their construction is explained as follows: —
Given an equation containing three variable quantities as

$$x = zy \qquad (194)$$

If we assume some value of z (making z a constant), the equation then becomes the equation of a right line.

If this line be platted, using rectangular coördinates (as the line $z = 1$ in the figure), then having given any value of y, the corresponding value of x may be taken off by scale. If a new value of z be assumed, the equation is obtained of a new line which may also be platted (as $z = \frac{1}{2}$ in the figure), and from which also, having given any value of y, the corresponding value of x may be determined by scale. Assuming a series of values of z and platting, we have a series of lines, each representing a different value of z, and from any one of which, having given a value of y, we may by scale determine the value of x.

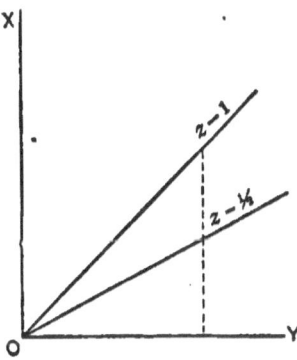

Thus, *given*, values of z and y; *required*, x, we may find,

1. The line corresponding to the given value of z, and
2. Upon this line we may find the value of x corresponding to the given value of y.

287. Next, if instead of platting upon *lines* as coördinate axes, we plat upon cross-section paper, the cross-section lines form a scale, so that the values of x and y need not be *scaled*, but may be *read* by simple inspection as in the figure.

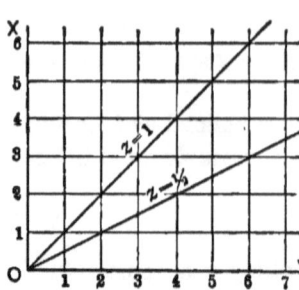

288. If the equation be in the form

$$x = azy \qquad (195)$$

the same procedure is equally possible, and the line representing any value of z will still be a right line.

If the equation be in the form

$$x = a(z+b)(y+c) + d \qquad (196)$$

in which a, b, c, d, are constants, the same procedure is still possible, and the line representing a given value of z is a right line, as before.

The use of diagrams of this sort is therefore possible for the solution of equations in the form of

$$x = a(z+b)(y+c) + d$$

or in simpler modifications of this form.

289. Referring again to the figure above, we may consider the horizontal lines to represent successive values of x and refer to them as the lines

$$x = 0\,;\ x = 1\,;\ x = 2,\ \text{etc.}$$

and similarly we may refer to vertical lines as the lines

$$y = 0\,;\ y = 1\,;\ y = 2,\ \text{etc.}$$

just as we refer to the inclined lines

$$z = \tfrac{1}{2}\,;\ z = 1,\ \text{etc.}$$

Having given any two of the quantities x, y, z, the third may be found by inspection from the diagram by a process similar to that described.

Earthwork Diagrams. 163

290. Diagram for Prismoidal Correction.

Formula $\qquad C = \dfrac{1}{3.24}(c_0 - c_1)(D_0 - D_1) \qquad$ (168)

This has the form $\quad x = \quad a \times \quad z \quad \times \quad y$

Construction of diagram.

Assume (as we did for z) a series of values of

$$c_0 - c_1 = 0, 1, 2, 3, 4, 5, \text{etc.}$$

When $c_0 - c_1 = 0$ then $C = 0$
or, the line $c_0 - c_1$ coincides with the line $C = 0$.

When $c_0 - c_1 = 1$, the equation of the line $c_0 - c_1$ is

$$C = \frac{1}{3.24}(D_0 - D_1)$$

To plat this right line, we must find two or more points on the line. For the reason that cross-section paper is generally warped somewhat, it is best to take a number of points not more than 3 or 4 inches apart, in order to get the lines sufficiently exact. For convenience, take values of $D_0 - D_1$ as follows : —
When $\qquad\qquad (c_0 - c_1) = 1$

take $D_0 - D_1 = 0, \quad 3.24, \quad 6.48, \quad 9.72, \quad 12.96, \quad 16.20,$ etc.

then $\qquad C = 0, \quad 1 \:,\quad 2 \:,\quad 3 \:,\quad 4 \:,\quad 5 \:,$ etc.

When $c_0 - c_1 = 2$, the equation of the line $c_0 - c_1$ is

$$C = \frac{1}{3.24} \cdot 2(D_0 - D_1)$$

Therefore when $\qquad c_0 - c_1 = 2$

take $D_0 - D_1 = 0, \quad 3.24, \quad 6.48, \quad 9.72, \quad 12.96, \quad 16.20,$ etc.

then $\qquad C = 0, \quad 2 \:,\quad 4 \:,\quad 6 \:,\quad 8 \:,\quad 10 \:,$ etc.

291.
In like manner a table may be constructed.

	0	3.24	6.48	9.72	12.96	16.20	19.44	22.68	26.92	D_0-D_1
0	0	0	0	0	0	0	0	0	0	
1	0	1	2	3	4	5	6	7	8	
2	0	2	4	6	8	10	12	14	16	
3	0	3	6	9	12	15	18	21	24	
4	0	4	8	12	16	20	24	28	32	
5	0	5	10	15	20	25	30	35	40	
6	0	6	12	18	24	30	36	42	48	
7	0	7	14	21	28	35	42	49	56	
8	0	8	16	24	32	40	48	56	64	
9	0	9	18	27	36	45	54	63	72	
10	0	10	20	30	40	50	60	70	80	
c_0-c_1										

292.
It will be noticed that when $D_0 - D_1 = 0$, $C = 0$.

Therefore for all values of $c_0 - c_1$, the lines pass through the origin.

We may proceed to plat the lines $c_0 - c_1 = 1$, $c_0 - c_1 = 2$, $c_0 - c_1 = 3$, etc., from data shown in the above table, platting upon the lines $D_0 - D_1 = 3.24$, $D_0 - D_1 = 6.48$, etc., the points shown with circles around them in the cross-section sheet, p. 165.

Having the lines $c_0 - c_1 = 1$, $c_0 - c_1 = 2$, 3, platted, intermediate lines are interpolated mechanically upon the principle that *vertical* lines would be proportionally divided (as ML is proportionally divided into 5 equal parts), and points are marked for the lines

$$c_0 - c_1 = 1.2, \quad 1.4, \quad 1.6, \quad 1.8$$

For the most convenient use, the values of $c_0 - c_1$ are taken to every second tenth of a foot in interpolating, as is shown on the diagram, p. 165, between 1 and 2; that is,

$$1.2, \quad 1.4, \quad 1.6, \quad 1.8$$

A complete diagram is shown at the back of the book.

Earthwork Diagrams.

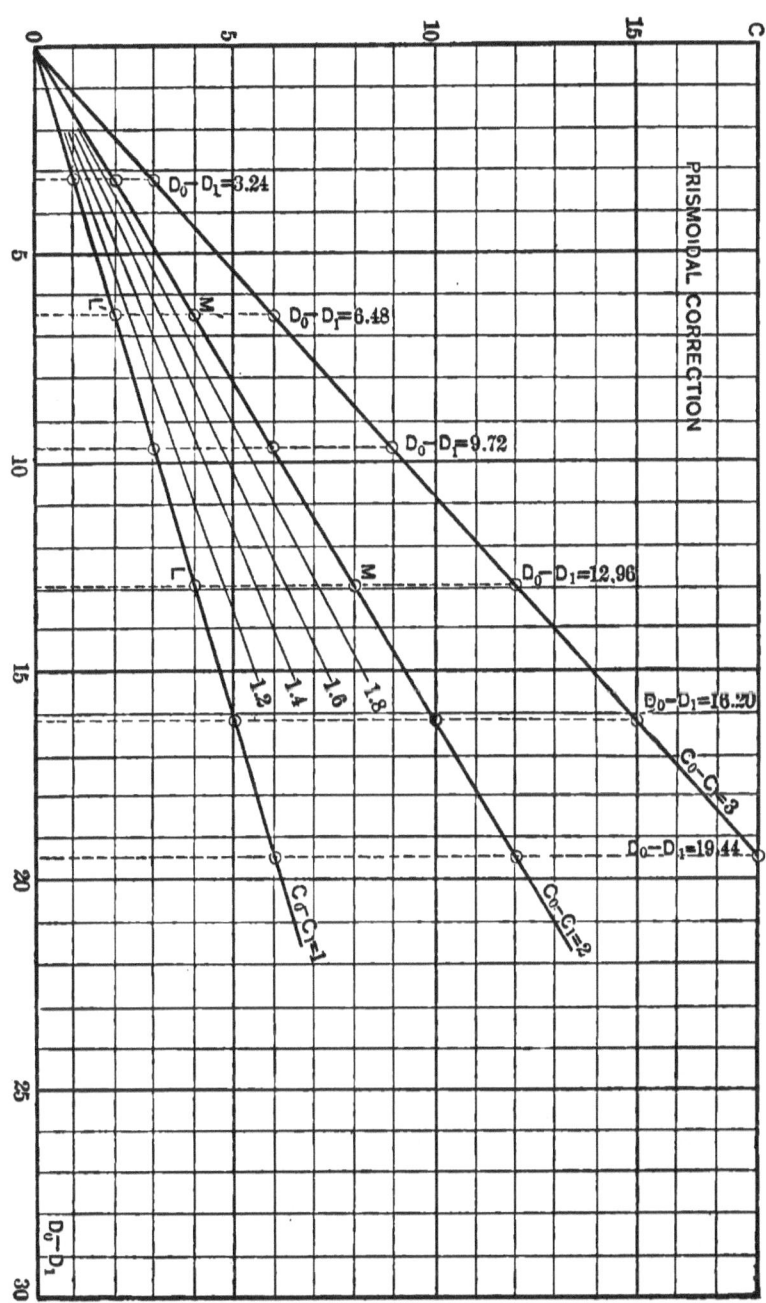

293. For Use.

Find the diagonal line corresponding to the given value of $c_0 - c_1$; follow this up until the vertical line representing the given value of $D_0 - D_1$ is reached, and the intersection is thus found. Then read off the value of C corresponding to this intersection.

Example. $c_0 - c_1 = 1.2$ $C = 4.0$
              $D_0 - D_1 = 11.0$

again, $c_0 - c_1 = 1.7$ $C = 3.6 \pm$
           $D_0 - D_1 = 7.0$

294. Diagram for Triangular Prisms.

From formula (191), $S = \dfrac{50}{54} cD$, a table may be constructed.

	0	5.4	10.8	16.2	21.6	27.0	D
0	0	0	0	0	0	0	
1	0	5	10	15	20	25	
2	0	10	20	30	40	50	
3	0	15	30	45	60	75	
4	0	20	40	60	80	100	
5	0	25	50	75	100	125	
6	0	30	60	90	120	150	
7	0	35	70	105	140	175	
8	0	40	80	120	160	200	
9	0	45	90	135	180	225	
10	0	50	100	150	200	250	
c							

From this a diagram can be constructed similar in form to that for Prismoidal Correction.

The lines for all values of c pass through the origin.

In constructing this table, any values of D might have been taken instead of those used here. Those used were selected because they give results simple in value, easily obtained, and readily platted.

Earthwork Diagrams.

295. Diagram for Three-Level Sections.

Formula, $\quad S = \dfrac{50}{54}\left(c + \dfrac{b}{2s}\right)D - \dfrac{50}{54} \cdot \dfrac{b}{2s} \cdot b \quad\quad (282)$

A separate diagram will be required for each value (or combination of values) of b and s. Since b and s thus become constants, the formula assumes the form of

$$x = a(z + b)y + d \quad\quad (197)$$

and the diagram will consist of a series of right lines.

A table can be made up by taking successive values of $c = 0$, 1, 2, 3, 4, etc., and finding for each of these the value of S corresponding to different values of D, using the above formula.

To make separate and complete computations directly by formula would be quite laborious; there is, however, a method of systematizing the construction of the *table* which can be shown better by example than in any other way.

296. Example. $\quad b = 14 \quad\quad s = 1\tfrac{1}{2} \text{ to } 1$

Formula $\quad S = \dfrac{50}{54}\left(c + \dfrac{b}{2s}\right)D - \dfrac{50}{54} \cdot \dfrac{b}{2s} \cdot b$

becomes $\quad S = \dfrac{50}{54}\left(c + \dfrac{14}{3}\right)D - \dfrac{50}{54} \cdot \dfrac{14}{3} \cdot 14$

$$S = \dfrac{50}{54}\left(c + \dfrac{14}{3}\right)D - 60.49 \quad\quad (198)$$

A table has been prepared for successive values of

$$c = 0, \quad 1, \quad 2, \quad 3, \quad 4, \quad 5, \quad \text{etc.}$$

and for $\quad D = 14, \quad 16.2, \quad 21.6, \quad 27.0, \quad \text{etc.}$

These values of D are selected for the following reasons: $D = 14$ is the least possible value; $D = 16.2, 21.6$ are desirable because they are multiples of 5.4, and the factors in the formula show that the computations will be simplified by selecting multiples of 5.4 for the successive values of D.

168 *Railroad Curves and Earthwork.*

	14	16.2	21.6	27.0	32.4	37.8	43.2	D
	12.963	15.	20.	25.	30.	35.	40.	Const. diff.
0	0	9.51	32.84	56.18	79.51	102.84	126.18	
1	12.963	24.51	52.84	81.18	109.51	137.84	166.18	
2	25.926	39.51	72.84	106.18	139.51	172.84	206.18	
3	38.889	54.51	92.84	131.18	169.51	207.84	246.18	
4	51.852	69.51	112.84	156.18	199.51	242.84	286.18	
5	64.815	84.51	132.84	181.18	229.51	277.84	326.18	
6	77.778	99.51	152.84	206.18	259.51	312.84	366.18	
7	90.741	114.51	172.84	231.18	289.51	347.84	406.18	
8	103.704	129.51	192.84	256.18	319.51	382.84	446.18	
9	116.667	144.51	212.84	281.18	349.51	417.84	486.18	
10	129.630	159.51	232.84	306.18	379.51	452.84	526.18	
c								

When $c = 0$ $S = \frac{50}{54} \cdot \frac{14}{3} \cdot D - 60.49$

When $D = 14$ $S = \frac{50}{54} \cdot \frac{14}{3} \cdot 14 - 60.49$

$\qquad\qquad\qquad = 60.49 - 60.49 = 0$

When $D = 16.2$

we may again calculate directly

$$S = \tfrac{50}{54} \cdot \tfrac{14}{3} \cdot 16.2 - 60.49$$

but a better method is to find how much greater S will be for $D = 16.2$ than for $D = 14.0$.

We have $S = \frac{50}{54} \cdot \frac{14}{3} \cdot D - 60.49$

Then for any new value D'

$$S' = \tfrac{50}{54} \cdot \tfrac{14}{3} \cdot D' - 60.49$$

$$S' - S = \tfrac{50}{54} \cdot \tfrac{14}{3}(D' - D) \qquad\qquad (199)$$

for $D' = 16.2$ $D = 14.0$ $D' - D = 2.2$

$$S' - S = \tfrac{50}{54} \cdot \tfrac{14}{3} \times 2.2 = 9.51$$

$$S = 0$$

$$S' = 9.51, \text{ which is entered in table.}$$

Earthwork Diagrams.

Similarly, $S'' - S' = \frac{50}{54} \cdot \frac{14}{3}(D'' - D')$

$D'' = 21.6 \qquad D' = 16.2 \qquad D'' - D' = 5.4$

$S'' - S' = \frac{50}{54} \times \frac{14}{3} \times 5.4$

$= 23.333$

$S' = 9.51 \qquad\qquad S^{iv} = 79.509$

$\overline{S'' = 32.843} \qquad\quad 23.333$

Similarly, $S''' - S'' = 23.333 \qquad \overline{S^v = 102.842}$

$\overline{S''' = 56.176} \qquad\quad 23.333$

$S^{iv} - S''' = 23.333 \qquad \overline{S^{vi} = 126.175}$

$\overline{S^{iv} = 79.509}$

Constant increment for $D' - D = 5.4$ is 23.333.

297. Each result is entered in the table in its proper place.
The final result for $c = 0$ and $D = 43.2$ should be calculated independently as a check.

When $\quad c = 0 \qquad S = \frac{50}{54} \cdot \frac{14}{3} \cdot D \quad - 60.49$

When $\quad D = 43.2 \qquad S = \frac{50}{54} \cdot \frac{14}{3} \times 43.2 - 60.49$

$\qquad\qquad\qquad\quad = 50 \times \frac{14}{3} \times 0.8 - 60.49$

$\qquad\qquad\qquad\quad = \frac{560}{3} \qquad\qquad - 60.49$

$\qquad\qquad\qquad\quad = 186.67 \qquad\quad - 60.49$

$\qquad\qquad\qquad\quad S = 126.18$

This checks exactly, and all intermediate values are checked by this process, which is also more rapid than an independent calculation for each value of D.

298. We now have values of S for the various values of $D = 14.0$, 16.2, 21.6, etc., when $c = 0$.

Next, find how much these will be increased when $c = 1$.

Formula $\qquad\qquad S = \frac{50}{54}(c + \frac{14}{3})D - 60.49$

for any new value $c' \qquad S' = \frac{50}{54}(c' + \frac{14}{3})D - 60.49$

$\qquad\qquad\qquad S' - S = \frac{50}{54}(c' - c)D \qquad\qquad (200)$

170 *Railroad Curves and Earthwork.*

When $c' = 1$ and $c = 0$, $\quad c' - c = 1$

$$S' - S = \tfrac{50}{54} D$$

Similarly, $\quad S'' - S' = \tfrac{50}{54}(c'' - c')D$

When $c'' = 2$ and $c' = 1$, $\quad c'' - c' = 1$

$$S'' - S' = \tfrac{50}{54} D$$

That is, for *any increase* of 1 ft. in the value of c,

$$S' - S = \tfrac{50}{54} D \tag{201}$$

When $\quad D = 14$

$$S' - S = \tfrac{50}{54} \times 14 = 12.963$$

This we enter as the constant difference for column $D = 14$.

We have already found $\quad S_0 = 0$
$\qquad\qquad\qquad\qquad\qquad\qquad\quad\ 12.963$
$\qquad\qquad\qquad\qquad\qquad S_1 = \overline{12.963}$
$\qquad\qquad\qquad\qquad\qquad\qquad\quad\ 12.963$
$\qquad\qquad\qquad\qquad\qquad S_2 = \overline{25.926}$

This gives column 14. $\qquad S_3 = 38.889$ etc.

When $\quad D = 16.2$

(201) $\quad S' - S = \tfrac{50}{54} D = \tfrac{50}{54} \times 16.2 = 50 \times 0.3$
$\qquad\qquad\qquad = 15$

Enter 15 as the constant difference in column 16.2.

We already have $\qquad S_0 = 9.51$
$\qquad\qquad\qquad\qquad\qquad\qquad\ \ 15.$
$\qquad\qquad\qquad\qquad\qquad S_1 = \overline{24.51}$
$\qquad\qquad\qquad\qquad\qquad\qquad\ \ 15.$
$\qquad\qquad\qquad\qquad\qquad S_2 = \overline{39.51}$

This allows us to complete column 16.2. $\quad S_3 = 54.51$ etc.

Similarly for $\quad D = 21.6 \quad\quad S' - S = 20$

Enter 20 as constant difference in column 21.6, and complete column as shown in table.

Similarly, fill out all the columns shown in the table.

Earthwork Diagrams.

299. The final result for $c = 10$, $D = 43.2$ should be calculated independently, and directly from the formula, as a check.

$$S = \tfrac{50}{54}(c + \tfrac{14}{3})D \quad - 60.49$$
$$c = 10 \quad D = 43.2$$
$$S = \tfrac{50}{54} \times 14.667 \times 43.2 - 60.49$$
$$= 50 \times 14.667 \times 0.8 - 60.49$$
$$= 40 \times 14.667 \quad - 60.49$$
$$= 586.68 \quad - 60.49$$
$$S = 526.19$$

The table gives 526.18. This checks sufficiently close to indicate that no error has been made. It would yield an exact check if we took $c + \tfrac{14}{3} = 14.6667$.

300. Note that for $\quad c = 10 \quad D = 43.2 \quad$ value $= 526.18$
$\qquad\qquad\qquad\qquad c = 10 \quad D = 37.8 \quad$ " $\quad 452.84$
$\qquad\qquad\qquad\qquad\qquad\qquad\qquad\qquad\qquad$ Diff. $= \overline{73.34}$

Between $\qquad c = 10 \quad D = 37.8 \;\}$ Diff. $= 73.33$
and $\qquad\quad\; c = 10 \quad D = 32.4 \;$

In line $c = 10$ a constant difference is found between successive values of D differing by 5.4. This may be demonstrated to be $= 73.33$.

All values in the table except column 14 are satisfactorily checked by applying this difference of 73.33 in line 10 together with the independent check of $c = 10$, $D = 43.2$.

The value of $c = 10$, $D = 14$ can also be checked and shown to be correct.

301. Having the table, page 168, completed, the construction of the diagram is simple.

The "*Diagram for Three-Level Sections, Base* 14, *Slope* 1½ *to* 1," was calculated and constructed according to this table. The Diagram given shows a general arrangement of lines and figures convenient for use. For rapid and convenient use, the diagram should be constructed upon cross-section paper, Plate G; and in this case the diagram will be upon a scale twice that of the diagram accompanying these notes.

172 Railroad Curves and Earthwork.

A "curve of level section" has been platted on this diagram in the following manner. For level sections, when

$c = 0$	$D = 14.0$	$c = 2$	$D = 20.0$
$c = 1$	$D = 17.0$	$c = 6$	$D = 32.0$
$c = 1.4$	$D = 18.2$ etc.		

The line passing through these points gives the "curve of level section."

Aside from the direct use of this curve of level section (for preliminary estimates or otherwise), it is very useful in tending to prevent any gross errors in the use of the table, since, in general, the points (intersections) used in the diagram will lie not far from the curve of level section.

302. Use of Diagram.

Find the diagonal line corresponding to the given value of c; follow this up until the vertical line representing the given value of D is reached, and this intersection found. Then read off the value of S corresponding to this intersection.

Example. Notes.

$$\text{Sta. 1} \quad \frac{13.0}{-4.0} \quad -3.7 \quad \frac{12.4}{-3.6} \quad S_1 = 136.$$

$$\text{Sta. 0} \quad \frac{10.6}{-2.4} \quad -2.5 \quad \frac{10.3}{-2.2} \quad \begin{array}{l} S_0 = 78. \\ S = 214. \end{array}$$

For Sta. 1 $c = 3.7$ $D = 25.4$

$c = 3.7$ is the middle of the space between 3.6 and 3.8.
Follow this up until the vertical line 25.4 is reached.
The intersection lies upon the line $S_1 = 136$.
Enter this above opposite Sta. 1.

For Sta. 0 $c = 2.5$ $D = 20.9$

$c = 2.5$ is the middle of space between 2.4 and 2.6.
Follow this up until the middle of space between 20.8 and 21.0 is reached.
The intersection lies just above the line
$$S_0 = 78$$
Enter this opposite Sta. 0.
$$S_{100} = S_1 + S_0$$
$$= 136 + 78 = 214 \text{ cu. yds.}$$

Earthwork Diagrams.

The prismoidal correction may be applied if desired.

It should be noticed that in each case the intersection was quite close to the "curve of level section."

303. Diagrams may be constructed in this way that will give results to a greater degree of precision than is warranted by the precision reached in taking the measurements on the ground.

In point of rapidity *diagrams are much more rapid than tables* for the computation of *Three-Level Sections.*

For "*Triangular Prisms*" and for *Prismoidal Correction*, the *diagrams* are *somewhat more rapid.*

For *Level Sections*, the tables for Three Level-Sections, § 274, are *at least equally rapid.*

A book entitled "Computation from Diagrams of Railway Earthwork," by Arthur M. Wellington, published by D. Appleton & Co., N.Y., explains the application and construction of certain other tables in addition to those given here. "Wellington's Diagrams," as there published, are upon a scale differing from that used here, and they do not allow of as great precision, but, on the other hand, are arranged to cover a large number of tables differing somewhat as to base and slope.

304. The use of approximate methods for applying the prismoidal correction to irregular sections (pp. 136–137) will be rendered practicable by the use of these "Diagrams for Three-Level Sections."

Method 1. No use of diagrams is necessary.

Method 2. Having found for any irregular sections (by triangular prisms or any other method) the solidity S for 50 ft. in length, find upon the diagram the line corresponding to this value of S; follow this line to the curve of level section, and read off the value of c (for a level section) which corresponds, and also the value of D for the same section.

Method 3. Having found in any way the value of S; if c is given, find the value of D to correspond; if D is given, find the value of c to correspond.

Method 4. The use of diagrams is not needed.

CHAPTER XVI.

HAUL.

305. When material from excavation is hauled to be placed in embankment, it is customary to pay to the contractor a certain sum for every cubic yard hauled. Oftentimes it is provided that no payment shall be made for material hauled less than a specified distance. In the east a common limit of "free haul" is 1000 ft. Often in the west 100 ft. is the limit of "free haul."

A common custom is to make the unit for payment of haul, one yard hauled 100 ft.; the price paid will often be from 1 to 2 cents per cubic yard hauled 100 ft.

The price paid for "haul" is small, and therefore the standard of precision in calculation need not be quite as fine as in the calculation of the quantities of earthwork. The total "haul" will be the product of

(1) the total amount of excavation hauled, and

(2) the average length of haul.

306. The average length of haul is the distance between the center of gravity of the material as found in excavation, and the center of gravity as deposited. It would not, in general, be simple to find the center of gravity of the entire mass of excavation hauled, and the most convenient way is to take each section of earthwork by itself. The "haul" for each section is the product of the

(1) number of cubic yards in that section, and

(2) distance between the center of gravity in excavation, and the center of gravity as deposited.

Haul. 175

307. When excavation is placed in embankment, there may be some difficulty in determining just where any given section of excavation will be placed, and where its center of gravity will be in embankment.

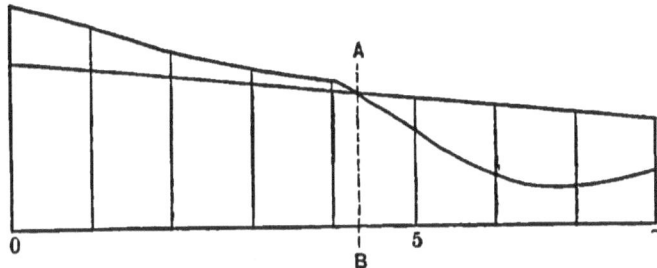

In hauling excavation in embankment, there is some plane, as indicated by AB, *to* which all excavation must be hauled on its way to be placed in embankment, and (another way of putting it) *from* which all material placed in embankment must be hauled on its way from excavation. We may figure the total haul as the sum of

(1) total "haul" of excavation *to* AB, and

(2) total "haul" of embankment *from* AB.

The total "haul" of excavation *to* AB and the total "haul" of embankment *from* AB will most conveniently be calculated as the sum of the hauls of the several sections of earthwork. For each section the haul is the product of

(1) the solidity S of that section, and

(2) distance from center of gravity of that section to the plane AB.

308. When the two end areas are equal, the center of gravity will be midway between the two end planes. When the two end areas are not equal in value, the center of gravity of the section will be at a certain distance from the mid-section, as shown by the formula

$$x_g = \frac{l^2}{12} \cdot \frac{A_1 - A_0}{S}$$

in which x_g = distance center of gravity from mid-section.

309. Referring to the figure below, and following the same general method of demonstration used on page 230, § 235,

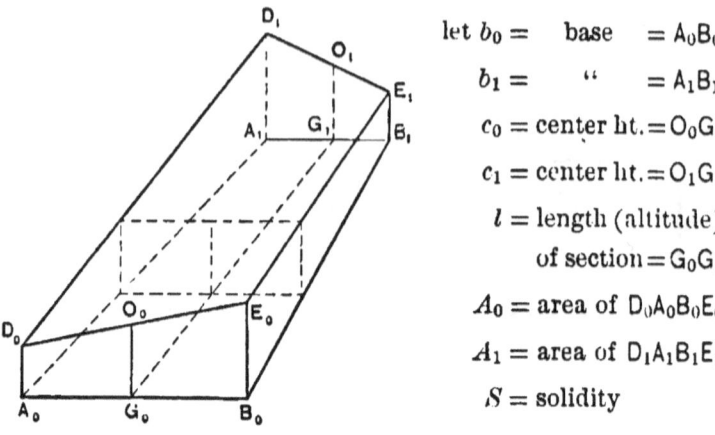

let b_0 = base = A_0B_0
b_1 = " = A_1B_1
c_0 = center ht. = O_0G_0
c_1 = center ht. = O_1G_1
l = length (altitude) of section = G_0G_1
A_0 = area of $D_0A_0B_0E_0$
A_1 = area of $D_1A_1B_1E_1$
S = solidity

Also use notation b_x, c_x, A_x for a section distant x from G_1.

Find the distance of the center of gravity from $A_1B_1E_1D_1$, and let this = x_c. Let x_g = distance of center of gravity from mid-section.

Then for any elementary section of thickness dx and distance x from $A_1B_1E_1D_1$, its moment will be

$$\left[b_1 + (b_0 - b_1)\frac{x}{l}\right]\left[c_1 + (c_0 - c_1)\frac{x}{l}\right] x\, dx$$

$$S \cdot x_c = \int_0^l \left[b_1 + (b_0 - b_1)\frac{x}{l}\right]\left[c_1 + (c_0 - c_1)\frac{x}{l}\right] x\, dx$$

$$S \cdot x_c = \frac{b_1 c_1 l^2}{2} + \frac{b_1(c_0 - c_1)l^3}{3l} + \frac{c_1(b_0 - b_1)l^3}{3l} + \frac{(c_0 - c_1)(b_0 - b_1)l^4}{4l^2}$$

$$= \frac{l^2}{12}\begin{bmatrix} 6 b_1 c_1 + 4 b_1 c_0 + 4 b_0 c_1 + 3 b_0 c_0 \\ -4 b_1 c_1 - 3 b_1 c_0 - 3 b_0 c_1 \\ -4 b_1 c_1 \\ +3 b_1 c_1 \end{bmatrix}$$

$$S \cdot x_c = \frac{l^2}{12} \times (b_1 c_1 + b_1 c_0 + b_0 c_1 + 3 b_0 c_0)$$

$$x_c = \frac{l^2}{12} \times \frac{b_1 c_1 + b_1 c_0 + b_0 c_1 + 3 b_0 c_0}{S} \qquad (202)$$

Haul. 177

What is wanted is x_g rather than x_c.

$$x_g = \frac{l}{2} - x_c$$

$$Sx_g = S\frac{l}{2} - Sx_c$$

$$S = \frac{l}{6}(2 b_1 c_1 + 2 b_0 c_0 + b_1 c_0 + b_0 c_1) \tag{164}$$

$$S \cdot \frac{l}{2} = \frac{l^2}{12}(2 b_1 c_1 + 2 b_0 c_0 + b_1 c_0 + b_0 c_1) \tag{203}$$

$$S \cdot x_c = \frac{l^2}{12}(b_1 c_1 + 3 b_0 c_0 + b_1 c_0 + b_0 c_1)$$

$$S \cdot x_g = \frac{l^2}{12}(b_1 c_1 - b_0 c_0)$$

$$= \frac{l^2}{12}(A_1 - A_0)$$

$$x_g = \frac{l^2}{12} \cdot \frac{A_1 - A_0}{S} \quad (S \text{ in cu. ft.}) \tag{204}$$

$$x_g = \frac{l^2}{12 \times 27} \cdot \frac{A_1 - A_0}{S} \quad (S \text{ in cu. yds.}) \tag{205}$$

310. The formula above applies to the solid shown in the figure, which has trapezoidal ends, but it will apply also when $D_0 A_0$, $D_1 A_1$, are each $= 0$, and therefore applies to such solids having triangular ends; and since any section of earthwork with parallel ends may be divided into a number of such solids with triangular ends, it applies to all ordinary sections of railroad earthwork, since it applies to the parts of which it is made up.

To show that in fact this formula is correct for prisms, wedges, and pyramids, use a method similar to that shown on page 129; find for each solid an expression for x_g in terms of A and l; reduce to the form

$$x_g = \frac{l^2}{12} \cdot \frac{A_1 - A_0}{S}$$

311. The formula

$$x_g = \frac{l^2}{12 \times 27} \cdot \frac{A_1 - A_0}{S}$$

is not in form convenient for use, because we have not found the values of A_1 and A_0, but instead have calculated directly from the tables or diagrams, the values of S_1 and S_0 for 50 ft. in length, where

$$S_1 = \frac{50}{27} A_1, \text{ or } A_1 = \frac{27\, S_1}{50}$$

and

$$A_0 = \frac{27}{50} S_0$$

Substituting, $\quad x_{g_{100}} = \dfrac{100 \times 100}{12 \times 27} \cdot \dfrac{S_1 - S_0}{S} \cdot \dfrac{27}{50}$

$$x_{g_{100}} = \frac{100}{6} \cdot \frac{S_1 - S_0}{S} \qquad (206)$$

This formula is in shape convenient for use, and results correct to the nearest foot can be calculated with rapidity.

312. For a section of length l less than 100 ft.

$$x_{g_l} = \frac{l^2}{12 \times 27} \cdot \frac{A_1 - A_0}{S_l}$$

$$= \frac{l^2}{12 \times 27} \cdot \frac{A_1 - A_0}{S_{100} \times \dfrac{l}{100}}$$

$$= \frac{100\, l}{12 \times 27} \cdot \frac{A_1 - A_0}{S_{100}}$$

$$x_{g_{100}} = \frac{100 \times 100}{12 \times 27} \cdot \frac{A_1 - A_0}{S_{100}}$$

$$x_{g_l} = x_{g_{100}} \cdot \frac{l}{100} \qquad (207)$$

Haul.

313. It is not, however, always necessary to calculate the position of the center of gravity of each station, or to calculate for each station the correction x_g. It may often be easy to calculate for a *series of sections* a correction to be applied to obtain the center of gravity of the entire mass.

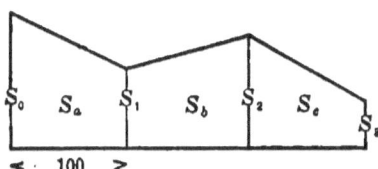

To find the position of the center of gravity of the entire mass, let

X_c = cent. of grav. for entire mass (approximately), using for each section c.g. at $\frac{l}{2}$

X = true dist. to c.g. of entire mass

$X_g = X - X_c$

$S_0 = \frac{50}{27} A_0$, $S_1 = \frac{50}{27} A_1$, S_2 = etc., as taken from tables or diagrams.

When all sections are of uniform length $= l$ as in figure above,

$$X_c S = \frac{l}{2}(S_a \qquad + 3\, S_b \qquad + 5\, S_c)$$

$$X S = S_a\left(\frac{l}{2} + x_{ga}\right) + S_b\left(3\frac{l}{2} + x_{gb}\right) + S_c\left(5\frac{l}{2} + x_{gc}\right)$$

$$S(X - X_c) = S_a x_{ga} \qquad + S_b x_{gb} \qquad + S_c x_{gc}$$

$$= \frac{100}{6}\left[S_a \frac{S_0 - S_1}{S_a} + S_b \frac{S_1 - S_2}{S_b} + S_c \frac{S_2 - S_3}{S_c}\right]$$

$$S X_g = \frac{100}{6}(S_0 - S_3) \qquad \text{or, in general,}$$

$$X_g = \frac{100}{6} \cdot \frac{S_0 - S_n}{S} \qquad (208)$$

where S is the solidity of the entire mass.

CHAPTER XVII.

MASS DIAGRAM.

314. Many questions of "haul" may be very usefully treated by means of a graphical method, known to some as "Mass Leveling," in which is used a diagram sometimes called a "Mass Profile," but which will be referred to here as the "Mass Diagram."

The construction of the "Mass Diagram" will be more clearly understood from an example than from a general description.

Let us consider the earthwork shown by the profile on p. 182, consisting of alternate "cut" and "fill." To show the work of constructing the "diagram" in full, the quantities are calculated throughout, but for convenience and simplicity, "level sections" are used and prismoidal correction disregarded. In a case in actual practice, the solidities will have been calculated for the actual notes taken.

315. In the table, p. 181, the

1st column gives the station.
2d column gives center heights.
3d column gives values of S from tables.
4th column gives values of S_{100} or S_l for each section, and with sign $+$ for cut or $-$ for fill.
5th column gives the total, or the sum of solidities up to each station; and in getting this total, each $+$ solidity is added and each $-$ solidity is subtracted, as appears in the table from the results obtained.

Having completed the table, the next step is the construction of the "Mass Diagram," page 182. In the figure shown there, each station line is projected down, and the value from column 5, corresponding to each station, is platted to scale as an offset from the base line at that station, all $+$ quantities above the line, and all $-$ quantities below the line. The points thus found are joined, and the result is the "Mass Diagram."

Mass Diagram. 181

Station.	Center Heights.	Solidity for 50' due to Center Height given (Taken from Tables).	Solidity for Section.	Solidity Totals.
0	0	0		0
1	+ 1.7	71	+ 71	+ 71
2	+ 2.7	120	+ 191	+ 262
3	0	0	+.120	+ 382
4	− 3.3	116	− 116	+ 266
5	− 5.1	204	− 320	− 54
6	− 2.9	99	− 303	− 357
7	0	0	− 99	− 456
8	+ 2.4	105	+ 105	− 351
9	+ 4.5	223	+ 328	− 23
10	+ 2.5	110	+ 333	+ 310
11	0	0	+ 110	+ 420
12	− 3.0	103	− 103	+ 317
13	− 5.3	215	− 318	− 1
14	− 7.6	357	− 572	− 573
15	− 8.4	414	− 771	− 1344
16	− 4.3	163	− 577	− 1921
17	0	0	− 163	− 2084
18	+ 2.6	115	+ 115	− 1969
19	+ 3.6	169	+ 284	− 1685
20	+ 4.9	248	+ 417	− 1268
21	+ 6.7	373	+ 621	− 647
22	+ 7.5	434	+ 807	+ 160
23	+ 5.2	268	+ 702	+ 862
24	+ 2.4	105	+ 373	+ 1235
25	0	0	+ 105	+ 1340
26	− 3.6	129	− 129	+ 1211
27	− 6.0	256	− 385	+ 826
28	− 5.0	199	− 455	+ 371
29	− 2.6	86	− 285	+ 86
30	0	0	− 86	0

182 *Railroad Curves and Earthwork.*

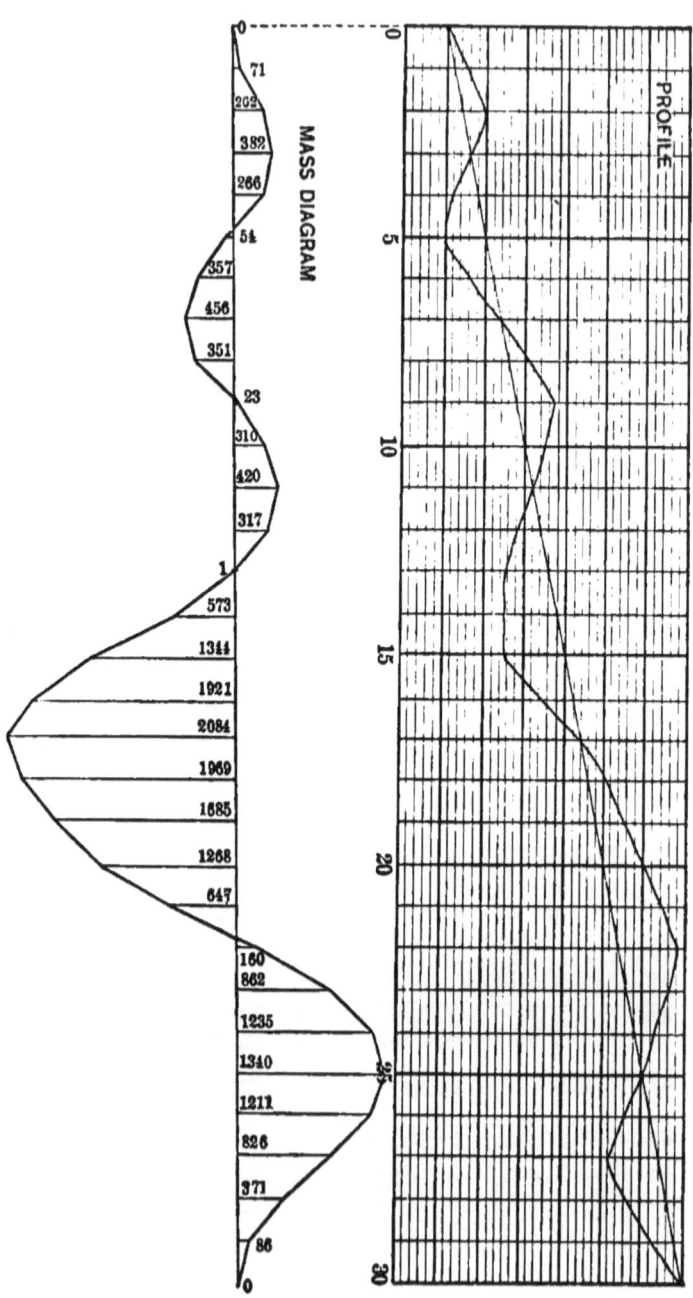

Mass Diagram. 183

316. It will follow, from the methods of calculation and construction used, that the "Mass Diagram" will have the following properties, which can be understood by reference to the profile and diagram, page 182.

1. Grade points of the profile correspond to maximum and minimum points of the diagram.
2. In the diagram, ascending lines mark excavation, and descending lines embankment.
3. The difference in length between any two vertical ordinates of the diagram is the solidity between the points (stations) at which the ordinates are erected.
4. Between any two points where the diagram is intersected by any horizontal line, excavation equals embankment.
5. The area cut off by any horizontal line is the measure of the "haul" between the two points cut by that line.

317. It may be necessary to explain the latter point at somewhat greater length.

Any quantity (such as dimension, weight, or volume) is often represented graphically by a line; in a similar way, the product of two quantities (such as volume into distance, or as foot pounds) may be represented or measured by an area. In the case of a figure other than a rectangle, the value, or product measured by this area, may be found by cutting up the area by lines, and these lines may be vertical lines representing volumes or horizontal lines representing distance. The result will be the same in either case. An example will illustrate.

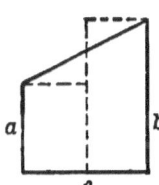 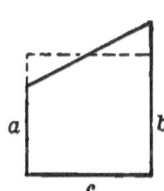

In the two figures let a and b represent pounds c " feet and the area of the trapezoid represent a certain number of foot pounds. The trapezoid may be resolved into rectangles by the use of a vertical line, as shown in Fig. 1, or by a horizontal line, as in Fig. 2.

In Fig. 1, the area is $a \times \dfrac{c}{2} + b \times \dfrac{c}{2}$

In Fig. 2, the area is $\dfrac{a+b}{2} \times c$

the result of course being the same in both cases.

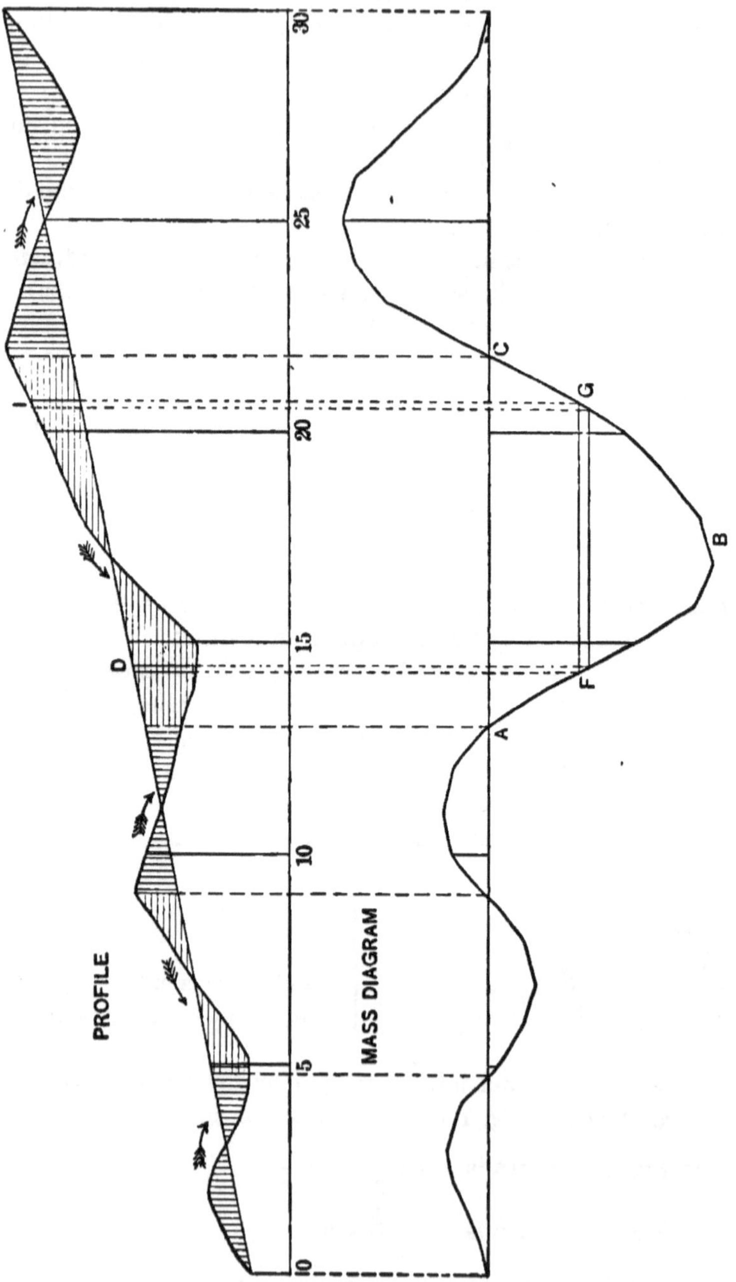

184   *Railroad Curves and Earthwork.*

318. In an entirely similar way, the area ABC (p. 184) represents the "haul" of earthwork (in cu. yds. moved 100 ft.) between A and C, and this area may be calculated by dividing it by a series of vertical lines representing solidities, as shown above G and F. That this area represents the haul between A and C may be shown as follows:—

Take any elementary solidity dS at D. Project this down upon the diagram at F, and draw the horizontal lines FG.

Between the points F and G (or between D and I), therefore, excavation equals embankment, and the mass dS must be hauled a distance FG, and the amount of "haul" on dS will be $dS \times$ FG, measured by the trapezoid FG. Similarly with any other elementary dS.

The total "haul" between A and C will be measured by the sum of the series of trapezoids, or by the area ABC. This area is probably most conveniently measured by the trapezoids formed by the vertical lines representing solidities. The average length of haul will be this area divided by the total solidity (represented in this case on p. 182 by the longest vertical line, 2084).

319. The construction of the "Mass Diagram" as a series of trapezoids involves the assumption that the center of gravity of a section of earthwork lies at its mid-section, which is only approximately correct since S for the first 50 ft. will seldom be exactly the same as S for the second 50 ft. of a section 100 ft. long. If the lines joining the ends of the vertical lines be made a curved line, the assumption becomes more closely accurate, and if the area be calculated by "Simpson's Rule," or by planimeter, results closely accurate will be reached.

It will be further noticed that hill sections in the "diagram" represent haul forward on the profile, and valley sections haul backward. The mass diagram may therefore be used to indicate the methods by which the work shall be performed; whether excavation at any point shall be hauled forward or backward; and, more particularly, to show the point where backward haul shall cease and forward haul begin, as indicated in the figure, p. 184, which shows a very simple case, the cuts and fills being evenly balanced, and no haul over 900 feet, with no necessity for either borrowing or wasting.

186   *Railroad Curves and Earthwork.*

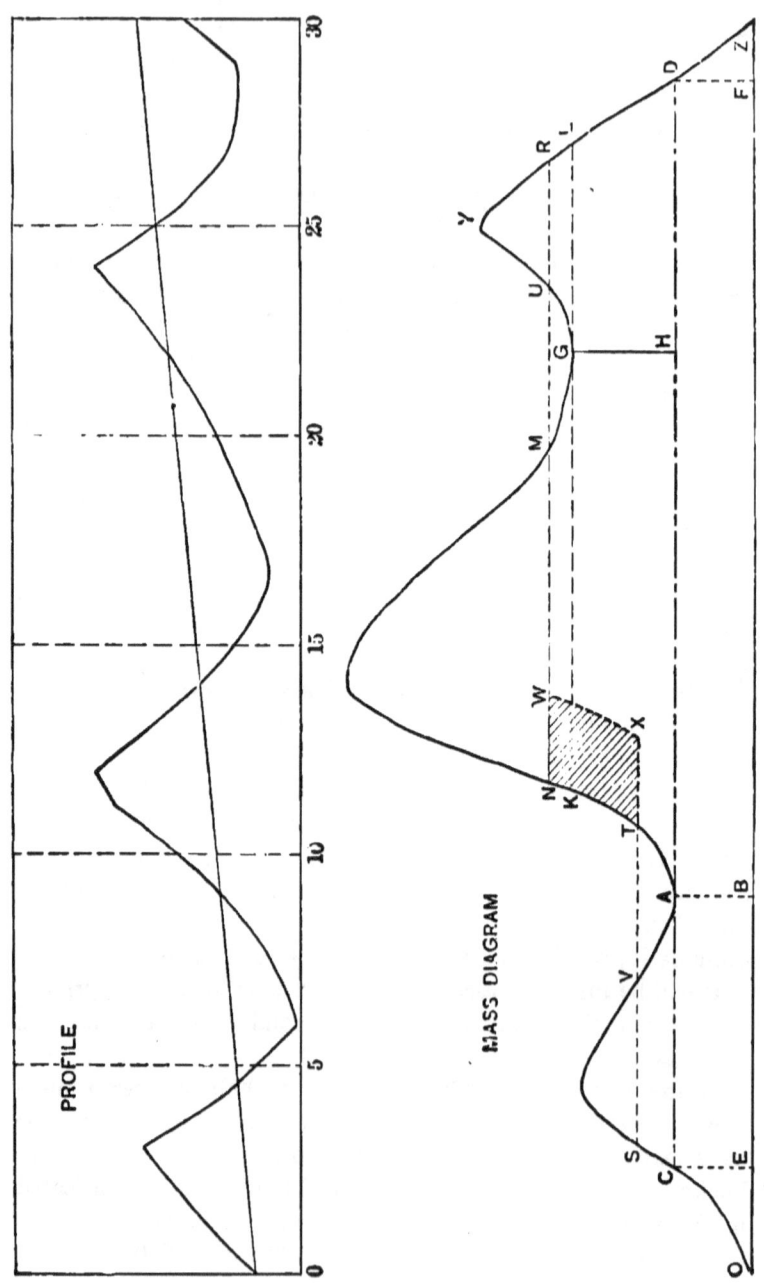

Mass Diagram. 187

320. In the figure, page 186, the excavation from Sta. 0 to 14 is very much in excess of embankment, and *vice versa* from Sta. 14 to 30. The mass diagram indicates a haul of nearly 3000 ft. for a large mass of earthwork, measured by the ordinate AB. It will not be economical to haul the material 3000 ft.; it is better to "waste" some of the material near Sta. 0, and to "borrow" some near Sta. 30, if this be possible, as is commonly the case.

If we draw the line CD, the cut and fill between C and D will still be equal, and the volume of cut measured by CE can be wasted, and the equal volume of fill measured by DF can be borrowed to advantage. It can be seen that there is still a haul of nearly 2000 ft. (from A to D) on the large mass of earthwork measured by GH. It is probable that it will not pay to haul the mass GH, or any part of it, as far as AD.

321. We must find the limit beyond which it is unprofitable to *haul* material rather than *borrow* and *waste*.

Let c = cost of 1 cu. yd. excavation or embankment.

h = cost of haul on 1 cu. yd. hauled 100 ft.

n = length of haul in "stations" of 100 ft. each.

Then, when 1 cu. yd. of excavation is wasted, and 1 cu. yd. of embankment is borrowed,

$$\text{cost} = 2c$$

When 1 cu. yd. of excavation is hauled into embankment,

$$\text{cost} = c + nh$$

The limit of profitable haul is reached when

$$2c = c + nh$$

or when
$$n = \frac{c}{h} \qquad (209)$$

Example. When excavation or embankment is 18 cents per cu. yd., and haul is 1½ cents,

$$n = \frac{18}{1.5} = 12 \text{ stations}$$

When $c = 16$ and $h = 2$

then $n = 8$ stations

188   *Railroad Curves and Earthwork.*

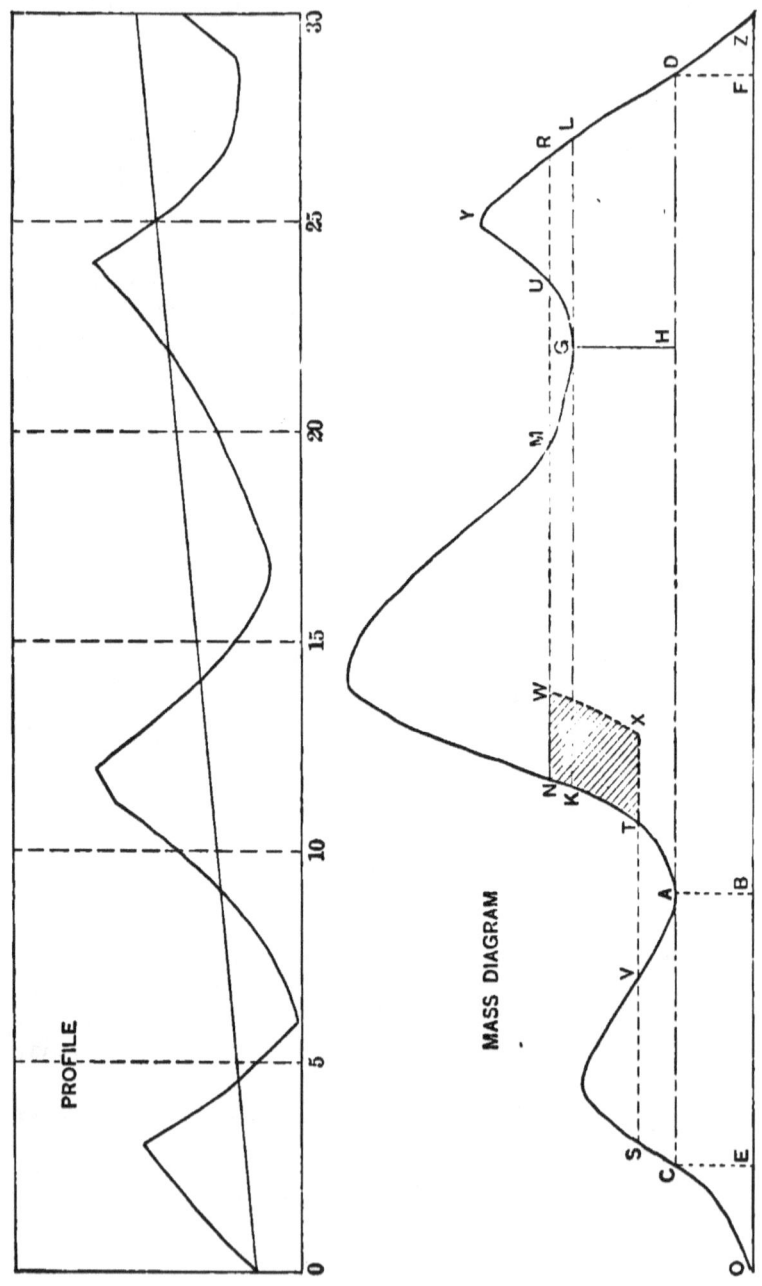

Mass Diagram.   189

322. In the former case (1200 ft. haul) we should draw in mass diagram (p. 188) the line KGL. Here KG is less than 1200 ft. The line should not be lower than G, for in that case the haul would be nearly as great as KL, or more than 1200 ft.

In the latter case (800 ft. haul) the line would be carried up to a point where NM = 800 ft. The masses between N and A, also C and O, can better be wasted than hauled, and the masses between M and G, also L and Z, can better be borrowed than hauled (always provided that there are suitable places at hand for borrowing and wasting).

Next, produce NM to R. The number of yards borrowed will be the same whether taken at RZ or at MG + LZ. That arrangement of work which gives the smallest "haul" (product of cu. yds. × distance hauled) is the best arrangement. The "haul" in one case is measured by GLRYU, and in the other by MGU + UYR. If MGU is less than GLRU, then it is cheaper to borrow (*a*) RZ rather than (*b*) MG + LZ.

In a similar way material NT and SO can be wasted more economically than NA and CO.

The most economical position for the line MR is when MU=UR. For ST, when SV = VT. For any change from these positions of MR and ST will show an increase of area representing "haul."

323. The case is often not as simple as that here given. Very often the material borrowed or wasted has to be hauled beyond the limit of "free haul." The limit beyond which it is unprofitable to haul will vary according to the length of haul on the borrowed or wasted material; the limit will, in general, be increased by the length of haul on the borrowed or wasted material. The haul on wasted or borrowed material, as NT, may be shown graphically by NTXW, where NW = TX shows the length of haul, and NTXW the "haul" (mass × distance).

The mass diagram can be used also for finding the limit of "free haul" on the profile, and various applications will suggest themselves to those who become familiar with its use and the principles of its construction. Certainly one of its most important uses is in allowing "haul" and "borrow and waste" to be studied by a diagram giving a comprehensive view of the whole situation. There are few if any other available methods of accomplishing this result.

Table for Three-Level Sections. Base 14', Slope 1½ to 1.

	0		.1		.2		.3		.4		
	L	K	L	K	L	K	L	K	L	K	
0	0.	6.5	2.6	6.6	5.3	6.8	8.0	6.9	10.8	7.0	0
1	28.7	7.9	31.9	8.0	35.1	8.1	38.4	8.3	41.7	8.4	1
2	63.0	9.3	66.7	9.4	70.5	9.5	74.3	9.7	78.2	9.8	2
3	102.8	10.6	107.1	10.8	111.4	10.9	115.8	11.1	120.3	11.2	3
4	148.1	12.0	153.0	12.2	157.9	12.3	162.8	12.5	167.9	12.6	4
5	199.1	13.4	204.5	13.6	209.9	13.7	215.4	13.8	221.0	14.0	5
6	255.6	14.8	261.5	15.0	267.5	15.1	273.6	15.2	279.7	15.4	6
7	317.6	16.2	324.1	16.3	330.7	16.5	337.3	16.6	344.0	16.8	7
8	385.2	17.6	392.2	17.7	399.4	17.9	406.5	18.0	413.8	18.1	8
9	458.3	19.0	466.0	19.1	473.6	19.3	481.4	19.4	489.1	19.5	9
10	537.0	20.4	545.2	20.5	553.4	20.6	561.7	20.8	570.1	20.9	10
11	621.3	21.8	630.0	21.9	638.8	22.0	647.7	22.2	656.6	22.3	11
12	711.1	23.1	720.4	23.3	729.7	23.4	739.1	23.6	748.6	23.7	12
13	806.5	24.5	816.3	24.7	826.2	24.8	836.2	25.0	846.2	25.1	13
14	907.4	25.9	917.8	26.1	928.3	26.2	938.8	26.3	949.3	26.5	14
15	1013.9	27.3	1024.8	27.5	1035.9	27.6	1046.9	27.7	1058.0	27.9	15
16	1125.9	28.7	1137.4	28.8	1149.0	29.0	1160.6	29.1	1172.3	29.3	16
17	1243.5	30.1	1255.6	30.2	1267.7	30.4	1279.9	30.5	1292.1	30.6	17
18	1366.7	31.5	1379.3	31.6	1392.0	31.8	1404.7	31.9	1417.5	32.0	18
19	1495.4	32.9	1508.5	33.0	1521.8	33.1	1535.1	33.3	1548.4	33.4	19
20	1629.6	34.3	1643.4	34.4	1657.1	34.5	1671.0	34.7	1684.9	34.8	20

	.5		.6		.7		.8		.9		
	L	K	L	K	L	K	L	K	L	K	
0	13.7	7.2	16.6	7.3	19.5	7.5	22.5	7.6	25.6	7.7	0
1	45.1	8.6	48.6	8.7	52.1	8.8	55.7	9.0	59.3	9.1	1
2	82.2	10.0	86.2	10.1	90.2	10.2	94.4	10.4	98.5	10.5	2
3	124.8	11.3	129.3	11.5	134.0	11.6	138.6	11.8	143.4	11.9	3
4	172.9	12.7	178.0	12.9	183.2	13.0	188.4	13.1	193.7	13.3	4
5	226.6	14.1	232.3	14.3	238.0	14.4	243.8	14.5	249.7	14.7	5
6	285.9	15.5	292.1	15.6	298.4	15.8	304.7	15.9	311.1	16.1	6
7	350.7	16.9	357.5	17.0	364.3	17.2	371.2	17.3	378.2	17.5	7
8	421.1	18.3	428.4	18.4	435.8	18.6	443.3	18.7	450.8	18.8	8
9	497.0	19.7	504.9	19.8	512.8	20.0	520.9	20.1	528.9	20.2	9
10	578.5	21.1	586.9	21.2	595.4	21.3	604.0	21.5	612.6	21.6	10
11	665.5	22.5	674.5	22.6	683.6	22.7	692.7	22.9	701.9	23.0	11
12	758.1	23.8	767.7	24.0	777.3	24.1	787.0	24.3	796.7	24.4	12
13	856.2	25.2	866.4	25.4	876.5	25.5	886.8	25.6	897.1	25.8	13
14	960.0	26.6	970.6	26.8	981.4	26.9	992.1	27.0	1003.0	27.2	14
15	1069.2	28.0	1080.4	28.1	1091.7	28.3	1103.1	28.4	1114.5	28.6	15
16	1184.0	29.4	1195.8	29.5	1207.7	29.7	1219.6	29.8	1231.5	30.0	16
17	1304.4	30.8	1316.7	30.9	1329.1	31.1	1341.6	31.2	1354.1	31.3	17
18	1430.3	32.2	1443.2	32.3	1456.2	32.5	1469.2	32.6	1482.2	32.7	18
19	1561.8	33.6	1575.3	33.7	1588.8	33.8	1602.3	34.0	1616.0	34.1	19
20	1693.8	35.0	1712.9	35.1	1726.9	35.2	1741.0	35.4	1755.2	35.5	20

Tables. 191

Table for **Three-Level Sections**. Base 20′, Slope 1½ to 1.

	0		.1		.2		.3		.4		
	L	K	L	K	L	K	L	K	L	K	
0	0.	9.3	3.7	9.4	7.5	9.5	11.4	9.7	15.3	9.8	0
1	39.8	10.6	44.1	10.8	48.4	10.9	52.8	11.1	57.3	11.2	1
2	85.2	12.0	90.0	12.2	94.9	12.3	99.9	12.5	104.9	12.6	2
3	136.1	13.4	141.5	13.6	147.0	13.7	152.5	13.8	158.0	14.0	3
4	192.6	14.8	198.5	15.0	204.6	15.1	210.6	15.2	216.7	15.4	4
5	254.6	16.2	261.1	16.3	267.7	16.5	274.3	16.6	281.0	16.8	5
6	322.2	17.6	329.3	17.7	336.4	17.9	343.6	18.0	350.8	18.1	6
7	395.4	19.0	403.0	19.1	410.7	19.3	418.4	19.4	426.2	19.5	7
8	474.1	20.4	482.2	20.5	490.5	20.6	498.8	20.8	507.1	20.9	8
9	558.3	21.8	567.1	21.9	575.9	22.0	584.7	22.2	593.6	22.3	9
10	648.1	23.1	657.4	23.3	666.8	23.4	676.2	23.6	685.6	23.7	10
11	743.5	24.5	753.4	24.7	763.3	24.8	773.2	25.0	783.2	25.1	11
12	844.4	25.9	854.8	26.1	865.3	26.2	875.8	26.3	886.4	26.5	12
13	950.9	27.3	961.9	27.5	972.9	27.6	984.0	27.7	995.1	27.9	13
14	1063.0	28.7	1074.5	28.8	1086.0	29.0	1097.7	29.1	1109.3	29.3	14
15	1180.6	30.1	1192.6	30.2	1204.7	30.4	1216.9	30.5	1229.1	30.6	15
16	1303.7	31.5	1316.3	31.6	1329.0	31.8	1341.7	31.9	1354.5	32.0	16
17	1432.4	32.9	1445.6	33.0	1458.8	33.1	1472.1	33.3	1485.4	33.4	17
18	1566.7	34.3	1580.4	34.4	1594.2	34.5	1608.0	34.7	1621.9	34.8	18
19	1706.5	35.6	1720.8	35.8	1735.1	35.9	1749.5	36.1	1764.0	36.2	19
20	1851.9	37.0	1866.7	37.2	1881.6	37.3	1896.5	37.5	1911.6	37.6	20
	0		.1		.2		.3		.4		

	.5		.6		.7		.8		.9		
	L	K	L	K	L	K	L	K	L	K	
0	19.2	10.0	23.2	10.1	27.3	10.2	31.4	10.4	35.6	10.5	0
1	61.8	11.3	66.4	11.5	71.0	11.6	75.7	11.8	80.4	11.9	1
2	110.0	12.7	115.1	12.9	120.2	13.0	125.5	13.1	130.8	13.3	2
3	163.7	14.1	169.3	14.3	175.1	14.4	180.9	14.5	186.7	14.7	3
4	222.9	15.5	229.1	15.6	235.4	15.8	241.8	15.9	248.2	16.1	4
5	287.7	16.9	294.5	17.0	301.4	17.2	308.3	17.3	315.2	17.5	5
6	358.1	18.3	365.4	18.4	372.8	18.6	380.3	18.7	387.8	18.8	6
7	434.0	19.7	441.9	19.8	449.9	20.0	457.9	20.1	466.0	20.2	7
8	515.5	21.1	524.0	21.2	532.5	21.3	541.0	21.5	549.7	21.6	8
9	602.5	22.5	611.6	22.6	620.6	22.7	629.7	22.9	638.9	23.0	9
10	695.1	24.1	704.7	24.0	714.3	—	724.0	24.3	733.7	24.4	10
11	793.3	25.2	803.4	25.4	813.6	25.5	823.8	25.6	834.1	25.8	11
12	897.0	26.6	907.7	26.8	918.4	26.9	929.2	27.0	940.0	27.2	12
13	1006.2	28.0	1017.5	28.1	1028.8	28.3	1040.1	28.4	1051.5	28.6	13
14	1121.1	29.4	1132.9	29.5	1144.7	29.7	1156.6	29.8	1168.5	30.0	14
15	1241.4	30.8	1253.8	30.9	1266.2	31.1	1278.6	31.2	1291.1	31.3	15
16	1367.4	32.2	1380.3	32.3	1393.2	32.5	1406.2	32.6	1419.3	32.7	16
17	1498.8	33.6	1512.3	33.7	1525.8	33.8	1539.4	34.0	1553.0	34.1	17
18	1635.9	35.0	1649.9	35.1	1664.0	35.2	1678.1	35.4	1692.2	35.5	18
19	1778.5	36.3	1793.0	36.5	1807.7	36.6	1822.3	36.8	1837.1	36.9	19
20	1926.6	37.7	1941.7	37.9	1956.9	38.0	1972.1	38.1	1987.4	38.3	20
	.5		.6		.7		.8		.9		

Table of Prismoidal Corrections.

c_0-c_1	1	2	3	4	5	6	7	8	9	c_0-c_1
D_0-D_1										D_0-D_1
0.1	.03	.06	.09	.12	.15	.19	.22	.25	.28	0.1
.2	.06	.12	.19	.25	.31	.37	.43	.49	.56	.2
.3	.09	.19	.28	.37	.46	.56	.65	.74	.83	.3
.4	.12	.25	.37	.49	.62	.74	.86	.99	1.11	.4
.5	.15	.31	.46	.62	.77	.93	1.08	1.23	1.39	.5
.6	.19	.37	.56	.74	.93	1.11	1.30	1.48	1.67	.6
.7	.22	.43	.65	.86	1.08	1.30	1.51	1.73	1.94	.7
.8	.25	.49	.74	.99	1.23	1.48	1.73	1.98	2.22	.8
.9	.28	.56	.83	1.11	1.39	1.67	1.94	2.22	2.50	.9
1.0	.31	.62	.93	1.23	1.54	1.85	2.16	2.47	2.78	1.0
.1	.34	.68	1.02	1.36	1.70	2.04	2.38	2.72	3.06	.1
.2	.37	.74	1.11	1.48	1.85	2.22	2.59	2.96	3.33	.2
.3	.40	.80	1.20	1.60	2.01	2.41	2.81	3.21	3.61	.3
.4	.43	.86	1.30	1.73	2.16	2.59	3.02	3.46	3.89	.4
.5	.46	.93	1.39	1.85	2.31	2.78	3.24	3.70	4.17	.5
.6	.49	.99	1.48	1.98	2.47	2.96	3.46	3.95	4.44	.6
.7	.52	1.05	1.57	2.10	2.62	3.15	3.67	4.20	4.72	.7
.8	.56	1.11	1.67	2.22	2.78	3.33	3.89	4.44	5.00	.8
.9	.59	1.17	1.76	2.35	2.93	3.52	4.10	4.69	5.28	.9
2.0	.62	1.23	1.85	2.47	3.09	3.70	4.32	4.94	5.56	2.0
.1	.65	1.30	1.94	2.59	3.24	3.89	4.54	5.19	5.83	.1
.2	.68	1.36	2.04	2.72	3.40	4.07	4.75	5.43	6.11	.2
.3	.71	1.42	2.13	2.84	3.55	4.26	4.97	5.68	6.39	.3
.4	.74	1.48	2.22	2.96	3.70	4.44	5.19	5.93	6.67	.4
.5	.77	1.54	2.31	3.09	3.86	4.63	5.40	6.17	6.94	.5
.6	.80	1.60	2.41	3.21	4.01	4.81	5.62	6.42	7.22	.6
.7	.83	1.67	2.50	3.33	4.17	5.00	5.83	6.67	7.50	.7
.8	.86	1.73	2.59	3.46	4.32	5.19	6.05	6.91	7.78	.8
.9	.90	1.79	2.69	3.58	4.48	5.37	6.27	7.16	8.06	.9
3.0	.93	1.85	2.78	3.70	4.63	5.56	6.48	7.41	8.33	3.0
.1	.96	1.91	2.87	3.83	4.78	5.74	6.70	7.65	8.61	.1
.2	.99	1.98	2.96	3.95	4.94	5.93	6.91	7.90	8.89	.2
.3	1.02	2.04	3.06	4.07	5.09	6.11	7.13	8.15	9.17	.3
.4	1.05	2.10	3.15	4.20	5.25	6.30	7.35	8.40	9.44	.4
.5	1.08	2.16	3.24	4.32	5.40	6.48	7.56	8.64	9.72	.5
.6	1.11	2.22	3.33	4.44	5.56	6.67	7.78	8.89	10.00	.6
.7	1.14	2.28	3.43	4.57	5.71	6.85	7.99	9.14	10.28	.7
.8	1.17	2.35	3.52	4.69	5.86	7.04	8.21	9.38	10.56	.8
.9	1.20	2.41	3.61	4.81	6.02	7.22	8.43	9.63	10.83	.9
4.0	1.23	2.47	3.70	4.94	6.17	7.41	8.64	9.88	11.11	4.0
.1	1.27	2.53	3.80	5.06	6.33	7.59	8.86	10.12	11.39	.1
.2	1.30	2.59	3.89	5.19	6.48	7.78	9.07	10.37	11.67	.2
.3	1.33	2.65	3.98	5.31	6.64	7.96	9.29	10.62	11.94	.3
.4	1.36	2.72	4.07	5.43	6.79	8.15	9.51	10.86	12.22	.4
.5	1.39	2.78	4.17	5.56	6.94	8.33	9.72	11.11	12.50	.5
.6	1.42	2.84	4.26	5.68	7.10	8.52	9.94	11.36	12.78	.6
.7	1.45	2.90	4.35	5.80	7.25	8.70	10.15	11.60	13.06	.7
.8	1.48	2.96	4.44	5.93	7.41	8.89	10.37	11.85	13.33	.8
.9	1.51	3.02	4.54	6.05	7.56	9.07	10.59	12.10	13.61	.9
5.0	1.54	3.09	4.63	6.17	7.72	9.26	10.80	12.35	13.89	5.0
c_0-c_1	1	2	3	4	5	6	7	8	9	c_0-c_1

Tables.

Table of Prismoidal Corrections.

c_0-c_1 D_0-D_1	1	2	3	4	5	6	7	8	9	c_0-c_1 D_0-D_1
5.1	1.57	3.15	4.72	6.30	7.87	9.44	11.02	12.59	14.17	5.1
.2	1.60	3.21	4.81	6.42	8.02	9.63	11.23	12.84	14.44	.2
.3	1.64	3.27	4.91	6.54	8.18	9.81	11.45	13.09	14.72	.3
.4	1.67	3.33	5.00	6.67	8.33	10.00	11.67	13.33	15.00	.4
.5	1.70	3.40	5.09	6.79	8.49	10.19	11.88	13.58	15.28	.5
.6	1.73	3.46	5.19	6.91	8.64	10.37	12.10	13.83	15.56	.6
.7	1.76	3.52	5.28	7.04	8.80	10.56	12.31	14.07	15.83	.7
.8	1.79	3.58	5.37	7.16	8.95	10.74	12.53	14.32	16.11	.8
.9	1.82	3.64	5.46	7.28	9.10	10.93	12.75	14.57	16.39	.9
6.0	1.85	3.70	5.56	7.41	9.26	11.11	12.96	14.81	16.67	6.0
.1	1.88	3.77	5.65	7.53	9.41	11.30	13.18	15.06	16.94	.1
.2	1.91	3.83	5.74	7.65	9.57	11.48	13.40	15.31	17.22	.2
.3	1.94	3.89	5.83	7.78	9.72	11.67	13.61	15.56	17.50	.3
.4	1.98	3.95	5.93	7.90	9.88	11.85	13.83	15.80	17.78	.4
.5	2.01	4.01	6.02	8.02	10.03	12.04	14.04	16.05	18.06	.5
.6	2.04	4.07	6.11	8.15	10.19	12.22	14.26	16.30	18.33	.6
.7	2.07	4.14	6.20	8.27	10.34	12.41	14.48	16.54	18.61	.7
.8	2.10	4.20	6.30	8.40	10.49	12.59	14.69	16.79	18.89	.8
.9	2.13	4.26	6.39	8.52	10.65	12.78	14.91	17.04	19.17	.9
7.0	2.16	4.32	6.48	8.64	10.80	12.96	15.12	17.28	19.44	7.0
.1	2.19	4.38	6.57	8.77	10.96	13.15	15.34	17.53	19.72	.1
.2	2.22	4.44	6.67	8.89	11.11	13.33	15.56	17.78	20.00	.2
.3	2.25	4.51	6.76	9.01	11.27	13.52	15.77	18.02	20.28	.3
.4	2.28	4.57	6.85	9.14	11.42	13.70	15.99	18.27	20.56	.4
.5	2.31	4.63	6.94	9.26	11.57	13.89	16.20	18.52	20.83	.5
.6	2.35	4.69	7.04	9.38	11.73	14.07	16.42	18.77	21.11	.6
.7	2.38	4.75	7.13	9.51	11.88	14.26	16.64	19.01	21.39	.7
.8	2.41	4.81	7.22	9.63	12.04	14.44	16.85	19.26	21.67	.8
.9	2.44	4.88	7.31	9.75	12.19	14.63	17.07	19.51	21.94	.9
8.0	2.47	4.94	7.41	9.88	12.35	14.81	17.28	19.75	22.22	8.0
.1	2.50	5.00	7.50	10.00	12.50	15.00	17.50	20.00	22.50	.1
.2	2.53	5.06	7.59	10.12	12.65	15.19	17.72	20.25	22.78	.2
.3	2.56	5.12	7.69	10.25	12.81	15.37	17.93	20.49	23.06	.3
.4	2.59	5.19	7.78	10.37	12.96	15.56	18.15	20.74	23.33	.4
.5	2.62	5.25	7.87	10.49	13.12	15.74	18.36	20.99	23.61	.5
.6	2.65	5.31	7.96	10.62	13.27	15.93	18.58	21.23	23.89	.6
.7	2.69	5.37	8.06	10.74	13.43	16.11	18.80	21.48	24.17	.7
.8	2.72	5.43	8.15	10.86	13.58	16.30	19.01	21.73	24.44	.8
.9	2.75	5.49	8.24	10.99	13.73	16.48	19.23	21.97	24.72	.9
9.0	2.78	5.56	8.33	11.11	13.89	16.67	19.44	22.22	25.00	9.0
.1	2.81	5.62	8.43	11.23	14.04	16.85	19.66	22.47	25.28	.1
.2	2.84	5.68	8.52	11.36	14.20	17.04	19.88	22.72	25.56	.2
.3	2.87	5.74	8.61	11.48	14.35	17.22	20.09	22.96	25.83	.3
.4	2.90	5.80	8.70	11.60	14.51	17.41	20.31	23.21	26.11	.4
.5	2.93	5.86	8.80	11.73	14.66	17.59	20.52	23.46	26.39	.5
.6	2.96	5.93	8.89	11.85	14.81	17.78	20.74	23.70	26.67	.6
.7	2.99	5.99	8.98	11.98	14.97	17.96	20.96	23.95	26.94	.7
.8	3.02	6.05	9.07	12.10	15.12	18.15	21.17	24.20	27.22	.8
.9	3.06	6.11	9.17	12.22	15.28	18.33	21.39	24.44	27.50	.9
10.0	3.09	6.17	9.26	12.35	15.43	18.52	21.60	24.69	27.78	10.0
c_0-c_1	1	2	3	4	5	6	7	8	9	c_0-c_1

ALLEN'S TABLES (Copyright, 1893, by C. F. ALLEN).
Triangular Prisms. S in cu. yds. for 50 ft. in length.

WIDTH	1	2	3	4	5	6	7	8	9	WIDTH
Height										*Height*
0.1	.09	.19	.28	.37	.46	.56	.65	.74	.83	0.1
.2	.19	.37	.56	.74	.93	1.11	1.30	1.48	1.67	.2
.3	.28	.56	.83	1.11	1.39	1.67	1.94	2.22	2.50	.3
.4	.37	.74	1.11	1.48	1.85	2.22	2.59	2.96	3.33	.4
.5	.46	.93	1.39	1.85	2.31	2.78	3.24	3.70	4.17	.5
.6	.56	1.11	1.67	2.22	2.78	3.33	3.89	4.44	5.00	.6
.7	.65	1.30	1.94	2.59	3.24	3.89	4.54	5.19	5.83	.7
.8	.74	1.48	2.22	2.96	3.70	4.44	5.19	5.93	6.67	.8
.9	.83	1.67	2.50	3.33	4.17	5.00	5.83	6.67	7.50	.9
1.0	.93	1.85	2.78	3.70	4.63	5.56	6.48	7.41	8.33	1.0
.1	1.02	2.04	3.06	4.07	5.09	6.11	7.13	8.15	9.17	.1
.2	1.11	2.22	3.33	4.44	5.56	6.67	7.78	8.89	10.00	.2
.3	1.20	2.41	3.61	4.81	6.02	7.22	8.43	9.63	10.83	.3
.4	1.30	2.59	3.89	5.19	6.48	7.78	9.07	10.37	11.67	.4
.5	1.39	2.78	4.17	5.56	6.94	8.33	9.72	11.11	12.50	.5
.6	1.48	2.96	4.44	5.93	7.41	8.89	10.37	11.85	13.33	.6
.7	1.57	3.15	4.72	6.30	7.87	9.44	11.02	12.59	14.17	.7
.8	1.67	3.33	5.00	6.67	8.33	10.00	11.67	13.33	15.00	.8
.9	1.76	3.52	5.28	7.04	8.80	10.56	12.31	14.07	15.83	.9
2.0	1.85	3.70	5.56	7.41	9.26	11.11	12.96	14.81	16.67	2.0
.1	1.94	3.89	5.83	7.78	9.72	11.67	13.61	15.56	17.50	.1
.2	2.04	4.07	6.11	8.15	10.19	12.22	14.26	16.30	18.33	.2
.3	2.13	4.26	6.39	8.52	10.65	12.78	14.91	17.04	19.17	.3
.4	2.22	4.44	6.67	8.89	11.11	13.33	15.56	17.78	20.00	.4
.5	2.31	4.63	6.94	9.26	11.57	13.89	16.20	18.52	20.83	.5
.6	2.41	4.81	7.22	9.63	12.04	14.44	16.85	19.26	21.67	.6
.7	2.50	5.00	7.50	10.00	12.50	15.00	17.50	20.00	22.50	.7
.8	2.59	5.19	7.78	10.37	12.96	15.56	18.15	20.74	23.33	.8
.9	2.69	5.37	8.06	10.74	13.43	16.11	18.80	21.48	24.17	.9
3.0	2.78	5.56	8.33	11.11	13.89	16.67	19.44	22.22	25.00	3.0
.1	2.87	5.74	8.61	11.48	14.35	17.22	20.09	22.96	25.83	.1
.2	2.96	5.93	8.89	11.85	14.81	17.78	20.74	23.70	26.67	.2
.3	3.06	6.11	9.17	12.22	15.28	18.33	21.39	24.44	27.50	.3
.4	3.15	6.30	9.44	12.59	15.74	18.89	22.04	25.19	28.33	.4
.5	3.24	6.48	9.72	12.96	16.20	19.44	22.69	25.93	29.17	.5
.6	3.33	6.67	10.00	13.33	16.67	20.00	23.33	26.67	30.00	.6
.7	3.43	6.85	10.28	13.70	17.13	20.56	23.98	27.41	30.83	.7
.8	3.52	7.04	10.56	14.07	17.59	21.11	24.63	28.15	31.67	.8
.9	3.61	7.22	10.83	14.44	18.06	21.67	25.28	28.89	32.50	.9
4.0	3.70	7.41	11.11	14.81	18.52	22.22	25.93	29.63	33.33	4.0
.1	3.80	7.59	11.39	15.19	18.98	22.78	26.57	30.37	34.17	.1
.2	3.89	7.78	11.67	15.56	19.44	23.33	27.22	31.11	35.00	.2
.3	3.98	7.96	11.94	15.93	19.91	23.89	27.87	31.85	35.83	.3
.4	4.07	8.15	12.22	16.30	20.37	24.44	28.52	32.59	36.67	.4
.5	4.17	8.33	12.50	16.67	20.83	25.00	29.17	33.33	37.50	.5

70

60

Diagram for
PRISMOIDAL CORRECTION

Differences between Sum of Distances out on Vertical Lines

Differences between Center Heights on Oblique Lines

Quantities on Horizontal Lines in cu. yds. for 100 ft. of Length

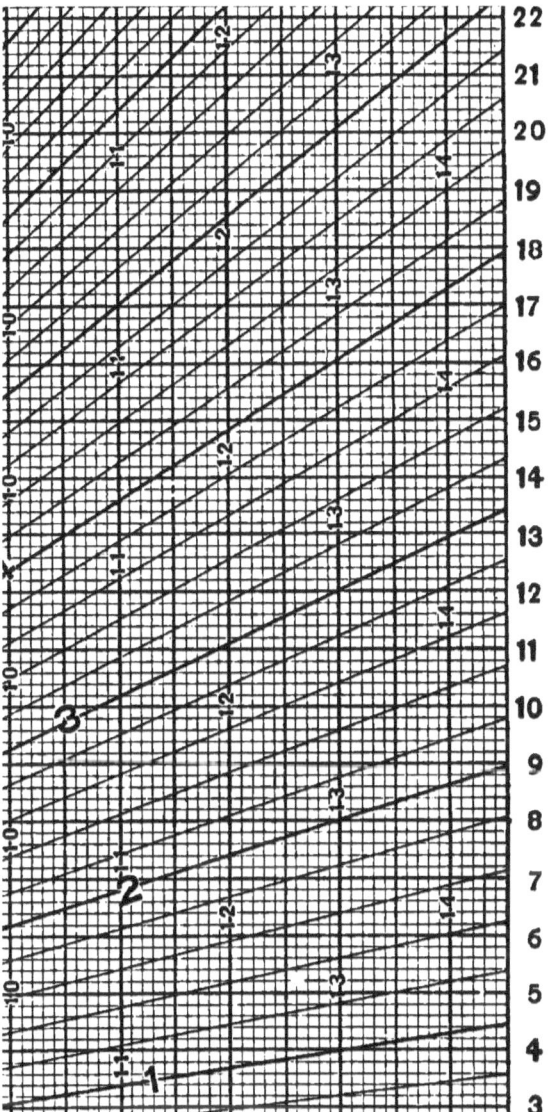

Diagram for
THREE LEVEL SECTIONS
Base 20. Slope 1½ to 1

Center Heights on Oblique Lines
Sum of Distances out on Vertical Lines
Quantities on Horizontal Lines in cubic
yards for 50 ft. of Length

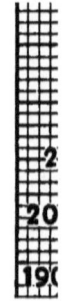

Diagram for
THREE LEVEL SECTIONS
Base 14. Slope 1½ to 1

Center Heights on Oblique Lines
Sum of Distances Out on Vertical Lines
Quantities on Horizontal Lines in
cubic yards for 50 ft. of Length

www.ingramcontent.com/pod-product-compliance
Lightning Source LLC
Chambersburg PA
CBHW031816230426
43669CB00009B/1165